Bayesian Compendium

Marcel van Oijen

Bayesian Compendium

Second Edition

Springer

Marcel van Oijen
Edinburgh, UK

ISBN 978-3-031-66087-0 ISBN 978-3-031-66085-6 (eBook)
https://doi.org/10.1007/978-3-031-66085-6

This Springer imprint is published by the registered company Springer Nature Switzerland AG
The registered company address is: Gewerbestrasse 11, 6330 Cham, Switzerland

Preface

Why This Book?

The recurring issue during my four decades in science has been a struggle with uncertainties. My primary field is agricultural, environmental and ecological science using both process-based and statistical models. And in each of my studies, uncertainties in data and models kept cropping up. I found that the literature contains many ways for classifying and analysing these uncertainties, but they mostly seem based on arbitrary criteria: a confusing menagerie of methods.

My encounters with probability theory, and with Bayesian methods in particular, showed that there is no need for confusion; there is a general perspective that can be applied in every case. The online papers by Edwin Jaynes, and his posthumously published book (2003), followed by the more practice-oriented small book by Sivia and Skilling (2006) did much to clarify matters. Very striking was Jaynes' explanation of the Cox postulates, which proved that consistent rational thinking requires the use of the rules of probability theory. Jaynes and Sivia took many of their examples from physics, but it became clear that Bayesian probabilistic methods are completely generic and can be applied in any field. But the methods do come with some practical problems. They require researchers to think carefully about their prior knowledge before embarking on an analysis. And the Bayesian analysis itself tends to be computationally demanding. Much research effort has gone into resolving these issues, and good methods are available.

So I have happily employed Bayesian methods during the past two decades. However, when teaching the approach to others, I found that there was no short yet comprehensive introductory text for scientists who do not have much statistical background. Especially amongst dynamic modellers, the study of statistical methods is not given high priority. But this book is intended to provide a quick and easy entry into Bayesian methods to all modellers, irrespective of the types of models and data that they work with. All chapters in this book are short, so it is more a compendium than a detailed manual.

Who Is This Book For?

This book is primarily for scientists who are newcomers to Bayesian methods, and who want to use data to parameterise and compare their models while accounting for uncertainties in data, model parameters and model structure. There are of course excellent books on Bayesian methods for beginners, but they tend to be very long texts that neglect the dynamic modeller. There is a need for short and easily digestible descriptions of Bayesian methods that cover the wide range of process-based and empirical models that are in common use.

The book is for everyone who works with models and data. Although aimed at beginners, I hope that more advanced users of Bayesian methods will also find material of interest. They may be unfamiliar with some of the more specialised topics or find the book useful for teaching. Readers may also be interested in the way this book tries to clarify the terminological confusion that reigns in scientific modelling and in applied statistics.

What Is in the Book?

The book begins with a gentle introduction to scientific uncertainty and Bayesian thinking. Once the basic ideas are in place, it provides explanations of a very wide range of modelling methods in three languages: English, mathematics and R. I believe that a mixture of the three languages, complemented with figures, is the best way to teach quantitative science. You may find English too woolly, mathematics too compact or R-code too problem-specific. But together they may provide clear, reasonably short and practical explanations. Some exercises are provided as well, and the solutions are at the end of the book.

The code examples in this book are all short, and they mostly use basic R that can easily be understood even if you are used to programming in a different computer language. R does come with a wealth of optional add-ons, called R-packages, but those will be used only occasionally.

Almost every example makes use of very simple artificial datasets. I mostly leave out the messiness of real datasets because that is distracting and obscures the view of how the methods work. Likewise, I focus on brevity and simplicity of code, rather than on speed of execution. The aim is to explain the diversity of Bayesian computational techniques that you can find in the literature. The algorithms that we show, such as Markov Chain Monte Carlo sampling (MCMC), can be used with any kind of model and data. In fact, one of the main advantages of Bayesian statistical modelling is that all idiosyncrasies of data, such as missing data, can be accounted for.

Downloadable Computer Code

The R-code used in this book is freely available. That includes the code chunks displayed in the book itself and the 'hidden' code used to make the figures. You can download it all from the author's public GitHub repository: https://doi.org/10.5281/zenodo.11092851. The code for each chapter is self-contained, so it can be run without downloading code for earlier chapters. When you find some of the textual explanations unclear or are struggling with an exercise, you may benefit from studying the code.

What Is New in This 2nd Edition?

The essence of the book has not changed since the first edition. It has the same key message: use Bayes to account for uncertainty because it is universally applicable, easy to understand, albeit sometimes hard to implement! The introductory text in the first two chapters has been reorganised, and the material on risk analysis, decision theory and machine learning has been expanded. Most chapters have had minor revisions to explain ideas more clearly. Literature references have been updated throughout and many new exercises were added.

Outline of Chapters

This book has seven sections, with the following contents:

- *The basics*: Chaps. 1–5
  - These chapters introduce the basics of scientific uncertainty and Bayesian thinking. In terms of abstract concepts, there is not much to learn: the Bayesian approach boils down to assigning a prior distribution and likelihood function, after which Bayes' Theorem tells us how to derive the posterior distribution.
- *Practicalities*: Chaps. 6–13
  - How exactly do we estimate parameters or compare models? Practical application of Bayesian methods can be hard if our models or data are complicated. We then rely on computer algorithms like MCMC to find representative samples from the posterior distribution. How does this work and how do we use the results?

- *Advanced material*: Chaps. 14–19
  - These chapters introduce methods that help us when modelling complicated or poorly known systems. We also show how Bayesian thinking supports risk analysis and decision making.
- *Specific models*: Chaps. 20–24
  - The models that we discuss in these chapters are simple functions or combinations of simple functions, sometimes very many of them. Yet some of these models are highly versatile and widely applicable, even in situations where we do not know beforehand what the shape of our function of interest is!
- *Outlook*: Chap. 25
  - This chapter reviews the state of Bayesian modelling and identifies trends of future methodological development. We also give suggestions for further self-study.
- *Appendices*
  - In these appendices, we present the notation we use in this book and very briefly summarise relevant mathematics, probability theory and software. The final appendix provides solutions to the exercises.
- *Index, references*

Edinburgh, UK Marcel van Oijen

Acknowledgements

Science is a collaborative affair, and over the years I have discussed Bayesian methods with many people. Most of them know how grateful I am to them, so I will do some cluster analysis and thank people by group—mentioning just a few names explicitly. I thank the many statisticians in the Netherlands and the UK who taught me over the years, including Ron Smith and Jonathan Rougier who were co-authors on my first Bayesian paper. To the present day, my publications with a Bayesian or otherwise probabilistic focus have been with about 140 different co-authors: a large group of inspiring people! Further inspiration came from colleagues in projects Nitro-Europe, Carbo-Extreme, GREENHOUSE, MACSUR, UK-SCaPE, PRAFOR and various COST-Actions, and from the Cambridge gang of July 2019. I thank my colleagues in the UK, the Netherlands, Belgium, Germany, France, Portugal, Hungary, Norway and Finland who invited me to teach, lecture or work on Bayesian methods, and the many students who showed interest and provided feedback. Mats Höglind gave me the opportunity for a sabbatical year in Norway where I spent much time reading Edwin Jaynes. I enjoyed organising a Bayesian course in Edinburgh because of the happy collaboration with Lindsay Banin, Kate Searle, David Cameron and Peter Levy, and the enthusiasm of the participants. David and Peter have been my main discussion partners over the years, so my extra special thanks go to them. Mark Brewer, Florian Hartig and two anonymous reviewers provided helpful comments on the first edition of this book. I also want to thank Eva Hiripi, my editor at Springer, for her kindness and her expert work on this book and thc 2022 monograph with Mark Brewer on risk analysis and decision theory.

Finally I want to mention, with more than gratitude, my wife Netty.

Contents

Chapter 1
Science and Uncertainty

Bayesian methods deal with uncertainty, and uncertainties are everywhere. This chapter provides an introduction to errors and uncertainty in science. The next chapter will introduce Bayesian thinking itself, as applied probability theory.

1.1 Measuring, Modelling and Science: The Three Basic Equations

In science, we use models to help us learn from data. But we always work with incomplete theory and measurements that contain errors. The consequence of this is that we need to acknowledge uncertainties when trying to extract information from measurements using our models. No measurement, explanation or prediction is perfectly accurate and precise. And we never know exactly how wrong they are. This book is about how to best extract information from data using models, acknowledging all relevant uncertainties. Our overall approach to science is captured in three equations. The first is the *Measurement Equation*, which acknowledges observational error:

$$\begin{aligned} &\underline{\text{MEASUREMENT EQUATION:}} \\ &y = z + \epsilon_y, \end{aligned} \tag{1.1}$$

M. van Oijen, *Bayesian Compendium*,
https://doi.org/10.1007/978-3-031-66085-6_1

where y is the measured value, z is the true value, and ϵ_y is measurement error. The second equation is the *Modelling Equation*, and it expresses the presence of a different kind of error:

$$\frac{\text{MODELLING EQUATION:}}{z = f(x, \theta) + \epsilon_{\text{model}},} \tag{1.2}$$

where f is the model, x are the environmental conditions for which we run the model, θ are the model's parameter values, and the final term ϵ_{model} represents model error which arises because f, x and θ will all be wrong to some extent. All the terms in the aforementioned equations can be scalars (single values), but more commonly they are vectors (multi-dimensional). Depending on the discipline, and sometimes the individual author or problem, the environmental conditions x are variously referred to as drivers, covariates, predictors, boundary conditions, explanatory or independent variables. Their defining characteristic is that we treat them as given rather than try to estimate them.

When we combine the Measurement and Modelling equation, we arrive at the *Science Equation*:

$$\frac{\text{SCIENCE EQUATION:}}{y = f(x, \theta) + \epsilon_{\text{model}} + \epsilon_y.} \tag{1.3}$$

Science is the attempt to explain observations y using a model f where we try to minimise measurement error ϵ_y and model error ϵ_{model} so that the model can be used to make predictions about the real world (z). The Bayesian approach to science recognises and quantifies uncertainties about all six elements of the Science Equation: y, f, x, θ, ϵ_{model}, and ϵ_y. We shall return to these equations at various points in the book (and expand them), but the main thing to notice for now is that we recognise errors in both measurement and modelling, which makes it hard to find the best parameter values θ and the best model form f.

1.2 Terminological Confusion

Because this book focuses on uncertainties in measuring and modelling, we need to adopt language from probability theory and statistics. The problem is that many concepts have been given very confusing names. Perhaps the most confusing term for scientists is *random variable*. Take the commonly heard statement 'To Bayesians, every parameter is a random variable that obeys a specific probability distribution'. That statement is not strictly wrong, but it is highly misleading. It suggests that parameters can jump around, assume arbitrary values. But Bayesians do not believe that. They believe that all parameters have specific values that we, however, can never know with infinite precision. So a random variable is just a real-world property about which we have incomplete knowledge. Randomness says nothing about the property, it says something about the state of knowledge of the researcher. And that state of knowledge will differ between people, so different researchers will quantify their uncertainties with different probability distributions. We shall try to avoid terms like 'random variable' in this book and only talk about uncertainties.

Of course, terminology can never be perfect. This book combines work from mathematics, logic, probability theory & statistics, computer science & programming and the sciences. These disciplines often use the same terms with different meanings, and they often use different terms to refer to the same thing. Even the most basic terms such as *model*, *parameter* and *process* have different definitions in the different fields. The following list defines our use of these and other terms.

- *Error*, *residual*, *uncertainty*, *variability*. These are terms that have different meanings but are often confused (Van Oijen, 2017).
  - *Error* is the difference between an estimate (measured or modelled) and the true value. So in Eq. (1.1), the term ϵ_y represents measurement error, and in Eq. (1.2), ϵ_{model} represents modelling error.
  - A *residual* or *mismatch* is the difference between a measured and a modelled value. Referring to the same two equations, the residuals are defined as $y - f(x, \theta)$.
  - *Uncertainty* is having incomplete knowledge about a quantity. It can always be represented by a probability distribution, denoted as $p[\,]$. For example, $p[\epsilon_y]$ defines which values of measurement error we deem possible and with what probability. Say our instrument is accurate (unbiased) but has only 50% precision. Then we are highly uncertain about the error and might assign to ϵ_y a Gaussian (or 'normal') distribution with mean zero and standard deviation equal to half the observed value. We could write that as $p[\epsilon_y] = N[\epsilon_y; \mu = 0, \sigma = y/2]$, but we mostly use shorter notation in this book where we specify the variance rather than the standard deviation: $p[\epsilon_y] = N[0, y^2/4]$ or equivalently $\epsilon_y \sim N[0, y^2/4]$.
  - *Variability* is the presence of differences in a set of values. Variability can lead to sampling error of uncertain magnitude. Say, we randomly sample 100 trees from a large forest, measure stem diameter on each and collect the 100 values in vector y. The variance σ_y^2 is then a measure of the variability in that sample. If all measurements are very precise, then there will be little uncertainty about that variability. But if we use the sample mean as an estimate for the mean diameter of all trees in the forest, then there will be an unknown sampling error. In this example, the uncertainty about the sampling error may be represented well by a normal distribution with mean zero and variance equal to $\sigma_y^2/100$.
- *Model*, *input*, *output*. A *model* (Fig. 1.1) is a function that converts inputs to outputs. Typical inputs are *parameters* and independent variables. Inputs and outputs can be scalars (one-dimensional) or vectors (multi-dimensional). Independent

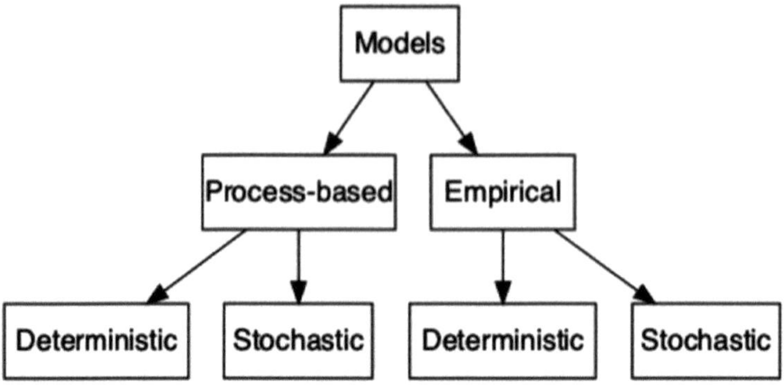

Fig. 1.1 The four model types. The deterministic models can be made stochastic by embedding them in a statistical model

variables can be time, spatial coordinates or environmental drivers such as time series of weather variables or human interventions to the modelled system. Models are used for two purposes: to explain the past and to predict the future. In both cases, the model is just a tool to extract information from measurements. Analysis of variance, for example, is a statistical model for extracting information from multi-factorial experiments. A complex global climate model is a tool to extrapolate past measurements of weather and other parts of the earth system to future climate change. There is no fundamental difference between the ANOVA and the GCM, both are imperfect data processors that model relationships in nature. The only difference is that the GCM is more complex.

  - *Dynamic model.* A model with time as a variable.
  - *Process-based model (PBM).* PBMs are deterministic dynamic models that simulate real-world processes. They typically have two types of input: drivers and parameters. Drivers are environmental variables (weather, disturbances, management), whereas parameters are constant properties of the system.
  - *Empirical model.* A model that does not dynamically simulate processes. So in this book, we use PBMs and empirical models as complementary terms. Every model belongs to either the first or the second category (Fig. 1.1).
  - *Deterministic model.* A model that produces numbers as output.
  - *Stochastic model.* A model that produces probability distributions as output. It is a complementary term to deterministic model: every model is either the one or the other (Fig. 1.1).
  - *Statistical model.* A stochastic model that explicitly represents uncertainty about errors by means of probability distributions.

- *Parameter.* A parameter is a constant input to a model. Examples of parameters are intercepts and slopes in empirical models and process rate constants in process-based models.
- *Prior, likelihood, posterior.* See Chap. 2.

- *Probability, random.* Randomness is in the eye of the beholder; believing it to be a property of the world is, in the words of Edwin Jaynes (2003), a 'mind-projection fallacy'. Likewise, probability is not a frequency but a state of mind. But it is fine to say that we sample a random value from a probability distribution (although even then we are ignoring the fact that a computer can only produce pseudo-random numbers). For example, we may want to sample from our prior distribution for the parameters of a model to see what that parameter uncertainty implies for predictive uncertainty.
- *Process.* In most cases, this will refer to a physical process in the real world. But it has a very different statistical meaning as a 'stochastic process', which is a probability distribution for the shape of a function. *Gaussian Processes* are an example of the latter, and they will feature quite prominently in this book (e.g. Chaps. 15 and 22). Both meanings of the term 'process' are so engrained in their fields that we will have to keep using both.

For now, these terms suffice, but we shall gradually be introducing more terminology.

1.3 Process-Based Models vs. Empirical Models

All methodology that we will present in this book can be applied to any kind of model: simple or complex, site-based or area-based, deterministic or stochastic, process-based or empirical. Most books that explain Bayesian methods use examples of small datasets and fairly simple models, and we shall in most cases do so too. It is a valid approach because the Bayesian thinking is the same in all cases. However, the complexity of a model does affect the ease of Bayesian computation. And, therefore, some of our examples will be drawn from the literature on deterministic process-based modelling of ecosystems. This is a class of dynamic models that simulates vegetation growth as part of the carbon-, water- and nutrient-cycles in the soil-tree-atmosphere system (Reyer, 2015). Such process-based models (PBMs) can be used for assessment of multiple ecosystem services (Mäkelä et al., 2012). PBMs tend to be parameter-rich and computationally demanding. We shall explain how these characteristics hamper the comprehensive application of the probabilistic approach to uncertainties and which solutions have been proposed. PBMs are particularly useful for the study of complex adaptive systems of which important internal processes and mechanisms are known, so we do not need to treat them as black boxes. PBMs aim to represent real mechanisms, i.e. causal relationships. You could say that empirical models do so too but only half-heartedly.

The literature only rarely shows statistical treatments of complex dynamic models such as PBMs. The applicability of the Bayesian perspective to such models may thus seem unclear. However, this book aims to show that the Bayesian perspective provides a unified treatment of all model-data issues, irrespective of the type of model or data.

1.4 Errors and Uncertainties in Modelling

The Bayesian approach is essential for scientists because it allows formal representation and comprehensive quantification of uncertainties as well as rigorous learning from observations. Errors are unavoidable in modelling, but we are uncertain about their magnitude. Inputs (parameters, environmental drivers) can never be known perfectly, and every model is a simplification of reality and thus has structural errors. We quantify our uncertainty about these errors in the form of probability distributions and propagate these uncertainties to model outputs, i.e. we assess how accurate and precise our model predictions are. We use data and Bayes' theorem to reduce the uncertainties about model error. To be able to use Bayesian methods, we need to have a clear understanding of the different kinds of error. So let's discuss them in detail.

1.4.1 Errors and Uncertainties in Model Drivers

As described earlier, drivers are the environmental conditions that are input to a model (the x in the Modelling Equation (1.2)). Methodologically, drivers are defined as boundary conditions that we do not attempt to calibrate, but that we accept from an external source of information. Although drivers are not calibrated, we are interested in any possible errors they may contain and their likely magnitude. Driver errors come in three types: gaps, measurement error and non-representativeness, which each need to be treated in a different way.

Gaps in driver data are common. When a model requires daily driver data, e.g. of weather conditions, and the weather time series misses a few days, the gaps need to be filled. Gap-filling methods range from simple linear interpolation to stochastic process modelling. The latter is the only approach for which uncertainties are readily quantified (as the so-called kriging or Gaussian process variance that we will meet again in Chap. 15).

Measurement error is caused by limited precision and accuracy of the measurement instrument, and sometimes also by errors in transferring measured values to data files. This is the simplest kind of error; uncertainty about it is often well represented as Gaussian noise.

The most complicated errors arise when the driver data are *not representative* of the environmental conditions for which we plan to run our model. The weather data may have come from an off-site weather station, or the data may have a different spatiotemporal scale than the model requires. A very common example is using output from gridded climate models as input for PBMs: the climatic data will then be grid cell averages, thus ignoring spatial heterogeneity. Such data are not representative for any specific ecosystem within the cell. If the model is nonlinear, as most models are, then input averaging leads to errors in model outputs, but methods to estimate that error exist, based on Taylor-expansion of the driver-dependency

(Van Oijen et al., 2017, see also Chap. 23). Another option is to downscale the data, although that brings its own errors and uncertainty. No general rule for assigning probability distributions to errors from non-representativeness exist: each case must be examined on its own.

1.4.2 Errors and Uncertainties in Model Parameters

Parameters are constants that represent fixed properties of our system of interest. In PBMs, every process that is modelled requires a parameter to define its basic rate. But process rates also depend on internal and external conditions (e.g. the nitrogen content of plants, or air temperature), and each controlling mechanism adds at least one sensitivity parameter to the model. The number of parameters in PBMs for vegetation modelling, for example, tends to range from several tenths to many hundreds (Cameron et al., 2013; Van Oijen et al., 2013b, 2008). This will pose problems for parameter estimation as we shall discuss later.

Parameters can be set to incorrect values, making model behaviour unrealistic. But a precise definition of parameter error is hard to give. What is the 'true' value of a parameter? Because a model is a simplification of reality, every parameter plays a somewhat different role in the model than its namesake does in the real world. We can therefore not simply go out and measure the true value exactly. This is apart from the fact that measurements have errors too and that no system property is ever truly constant in reality. In practice, we say that the correct value of a parameter is the value that makes the model behave most realistically. Parameter error is the difference from that value, and parameter uncertainty is not knowing what that value is.

Although measurements cannot provide true parameter values, they can give first rough estimates. For about 30 different parameters common to three forest PBMs Levy et al. (2005) reviewed measurement values reported in the literature. For each parameter, a wide range of values was reported that could be interpreted as probability distributions representing parameter uncertainty. For many parameters, the distributions were skewed and better represented by beta distributions than by normal distributions (Van Oijen & Thomson, 2010).

1.4.3 Errors and Uncertainties in Model Structure

Every model is a simplification of reality and therefore, to some extent, wrong. But the behaviour of models—their repertoire of outputs for different conditions—can be compared, showing that some models are more realistic than others. Comparisons of ecosystem models abound in the literature and tend to show large differences between their predictions (Cameron et al., 2013; Van Oijen et al., 2008). Some tentative general conclusions may be drawn: the feedback structure of models is more important than the mathematical form of individual equations (Hyvönen et al., 2007), and a consistent level of process detail in different parts of the model is desirable (Rollinson et al., 2017; Van Oijen et al., 2004).

Increasingly, comparisons of all models, including complex PBMs, involve assessing simulations against observations, not just of output variables such as productivity but also of the underlying processes and mechanisms as represented in the models (Clark et al., 2016; Hickler et al., 2015; Johnson et al., 2016; Schlesinger et al., 2016). Medlyn et al. (2015) refer to this as an 'assumption-centred model intercomparison approach'. However, these model comparisons do not employ probability theory and therefore cannot quantify the degree of uncertainty about model structural error. Moreover, any advice about model structure based on observations is contingent on the range of environmental conditions for which the models were tested and remains heuristic: there is no unique way to derive a model from first principles, so errors remain inevitable.

This leaves us with only two ways to account, probabilistically, for structural error in a modelling study: extend the PBM with a stochastic error term (Kennedy & O'Hagan, 2001), or use a large ensemble of different models and proceed as if one model in the set should be correct (e.g. Fu et al., 2012). In the first approach, uncertainty is quantified by assigning a probability distribution to the structural error term. In the second approach, uncertainty is represented by a probability distribution over the set of models, with highest probabilities assigned to models that are considered to be most plausible. Both require taking into account that model performance depends not only on its structure but also on the parameter settings with their own uncertainties. Technical details of both methods will be given in Chaps. 11 and 12.

1.4.4 Forward Propagation of Uncertainty to Outputs

Because measurements and models have unknown errors, all we can quantify are uncertainties. Therefore, the common term of 'error propagation' denoting how much the error in inputs contributes to error in outputs is a misnomer. What is propagated is uncertainty, not error. Outputs may have minimal error despite large errors in inputs, if the input errors happen to have compensating effects. This is a common occurrence when models are tuned to produce a desired result.

The techniques, rather than the name, associated with 'error propagation' may well be used to quantify output uncertainty, provided the model is simple enough that partial derivatives of output with respect to inputs can be analytically calculated. However, PBMs tend to be too complex for such approaches, so uncertainty is mostly quantified by Monte Carlo methods: sampling from the probability distributions of model inputs (and structures if we have an ensemble of models) to generate a representative set of possible model outputs. That set of outputs is a sample from the model's predictive distribution. In the study by Levy et al. (2005), aforementioned, Monte Carlo sampling was used to quantify the contribution of parameter and model structural uncertainty to uncertainty about the carbon sink of a coniferous forest in southern Sweden. They concluded that the carbon sink uncertainty was for 92% due to parameter uncertainty, whereas model structural uncertainty accounted for only 8%. The key parameter uncertainties were for allocation of carbon to leaves, stems and roots. These results were of course contingent on the three chosen forest models and the single application site. Reyer et al. (2016) also demonstrated the importance of parameter uncertainty—in their case for prediction of future forest productivity—but in a comparison with uncertainty about climatic drivers rather than model structure. Sutton et al. (2008) showed the importance of uncertainties about model drivers (in particular atmospheric nitrogen deposition) and model structure for predictions of forest productivity across Europe. Minunno et al. (2013a) found that uncertainty about soil conditions (water availability and fertility) mainly determined the predictive uncertainty of a growth model for *Eucalyptus globulus* in Portugal.

Let us now turn to a more thorough exploration of the use of probability theory in dealing with uncertainties, bringing us to the topic of *Bayesian inference*.

Exercise

1. **Uncertainty propagation.** Say that you have a model f with independent uncertain inputs: $x \sim N[\mu = 1, \sigma^2 = 1]$ and $\theta \sim N[\mu = 2, \sigma^2 = 1]$. What is the model's predictive uncertainty if: **a.** $f(x, \theta) = \theta + x$, **b.** $f(x, \theta) = \theta - x$, **c.** $f(x, \theta) = \theta x$, **d.** $f(x, \theta) = \exp\left(-(x + \theta)^2\right)$? You can ignore uncertainty about model structure for now. For models **a** and **b**, give a formula for the output distribution. For models **c** and **d**, write a one-line R-program that produces a sample ($n = 10{,}000$) from the output distribution. You may want to consult Appendix D to refresh your knowledge of R.

Chapter 2
Bayesian Inference

The preceding chapter discussed how scientific research is always carried out under conditions of incomplete information. We never know 'the whole story', much remains uncertain. Scientists therefore increasingly turn to probability theory when applying their models (e.g. Hartig et al., 2012; Jaynes, 2003; Ogle & Barber, 2008; Sivia & Skilling, 2006; Van Oijen & Brewer, 2022). This is the approach that we take in this book too. We show how defining all our uncertainties as probability distributions allows for rigorous reduction of those uncertainties when new data come in. The approach that we are presenting is known in the literature under many different names, including *Bayesian calibration*, *data assimilation*, *model-data fusion* and *inverse modelling*. Whilst the different names refer to different applications of modelling, they all share the idea of specifying probability distributions and modifying them according to the rules of probability theory (in particular as we shall see, *Bayes' Theorem*) when new data become available. This manner of drawing probabilistic conclusions is called *Bayesian inference*. But how does it work exactly?

2.1 Thinking Like a Bayesian

The basics of Bayesian inference are simple. There are just three elements, connected by probability theory. The elements are: (1) your prior belief about a quantity or proposition, (2) new information and (3) your posterior belief. Probability theory provides the logical connection from the first two elements to the last. So all we need to learn is how to express beliefs and new information in the form of probability distributions, and then we can simply follow the rules of probability theory. That is all!

M. van Oijen, *Bayesian Compendium*,
https://doi.org/10.1007/978-3-031-66085-6_2

While the simplicity of Bayesian inference is a given, that does not mean it is necessarily easy to learn. Consistently thinking in terms of probability theory is not second nature to everyone. And there is no unique simplest way to teach Bayesian thinking to you. It all depends on your background, and your favourite mode of rational thinking. Do you prefer to begin with abstract concepts, then mathematical equations, then examples? Or do you wish to begin with puzzles or anecdotes and learn how they can all be approached in the same way? Perhaps you like to start from your knowledge of classical statistics and learn how its methods can always be interpreted, and often improved, in a Bayesian way? But here we begin with a short detective story ...

2.2 A Murder Mystery

You are a detective, and you are called to a country house: the owner has been found murdered in the library. The three possible suspects are his wife, his son and the butler. *Before reading on, who do you believe committed the crime?* And do not say: 'I can't answer that, I have not inspected the evidence yet'. You are a Bayesian detective, so you can state your prior probabilities.

Your assistant says, 'I bet it's the wife', but you find that silly bias. You see the butler as the prime suspect and would give odds of 4:1 that he is the culprit. You find the wife just as improbable as the son. So your prior probability distribution for wife:son:butler is 10:10:80%. Of course, you would not really bet money on the outcome of the case—you're a professional—but you decide to investigate the butler first. To your surprise, you find that the butler has a perfect alibi. *What is your probability distribution now?* The answer is that the alibi of the butler has no bearing on the wife or son, so they remain equally likely candidates, and your probability distribution becomes 50:50:0%.

Next, you inspect the library and find that the murder was committed with a blunt instrument. *How does that change your probabilities?* You assess the likelihood of such a murder weapon being chosen by a man to be twice as high as by a woman. So that changes your probabilities to 33:67:0%.

I leave you to finish the story to a logical conclusion where sufficient evidence has been processed such that the murderer is identified beyond reasonable doubt. But what does this murder mystery teach us? Well, our story is fiction, but it contains the three steps of any Bayesian analysis: assigning a prior probability distribution, acquiring new information and updating your probability distribution following the rules of probability theory. Everything in this book, and in Bayesian statistics generally, is about one or more of these three steps. So if you found the detective's reasoning plausible, then you are already a Bayesian! In fact, there is a rich literature in the field of psychology that shows that human beings at least to some extent make decisions in a Bayesian way (Hahn, 2014; Rule et al., 2020).

2.3 Bayes' Theorem

Bayes' Theorem dates back to the eighteenth century (Bayes, 1763), and its use in science has been explained in textbooks (e.g. Gelman et al., 2013; McElreath, 2016) and tutorial papers (e.g. Hartig et al., 2012; Van Oijen et al., 2005). So what is the theorem? To state it, we need to introduce some notation (you can find more in the Appendices). We write $p[A|B]$ for the *conditional probability* of A being true given that B is true. The theorem is then written as follows:

BAYES' THEOREM:

$$p[A|B] = \frac{p[A]\,p[B|A]}{p[B]}. \tag{2.1}$$

All four terms in Eq. (2.1) are probabilities or probability densities, and the theorem is true irrespective of what is denoted by A and B. Statement A is what we are interested in. It can stand for 'it will rain tomorrow' or 'crop yield will exceed 10 tons per hectare' or 'the biomass density parameter has value 900 kg m^{-3}'. We want to quantify the probability that A is true given some information B, and we write this conditional probability as $p[A|B]$. In scientific applications, A often refers to a parameter or parameter vector whose true value θ we want to know, so A could stand for "$\theta = 1$" or "$\theta = 2$" or any other parameter value. And B could then be the statement 'my sensor gives me the value y'. Note that in such cases, Bayes' Theorem actually gives us infinitely many different correct probability statements at once, because it remains valid for every possible parameter value θ and any data y. In other words, Bayes' Theorem defines the whole conditional probability *distribution* for the parameter value given the data. Here is Bayes' Theorem again using these more common symbols:

BAYES' THEOREM (common notation):

$$p[\theta|y] = \frac{p[\theta]p[y|\theta]}{p[y]}. \tag{2.2}$$

All four terms in the theorem have a name. $p[\theta|y]$ is the *posterior probability distribution* for the parameter θ, $p[\theta]$ is its *prior distribution*, $p[y|\theta]$ is the *likelihood function*, and $p[y]$ is the *evidence*. Don't confuse the last term with the criminal evidence in our murder mystery! And be aware that the likelihood $p[y|\theta]$ is often written succinctly as $L[\theta]$. Your job in a Bayesian analysis will be to define the prior and likelihood, and the evidence will then automatically be defined as well (because total probability always equals 1), after which Bayes' Theorem gives us the posterior distribution that we are seeking. How to define the prior and likelihood will be explained in Chaps. 3 and 4.

A brief aside on notation. We just gave names to the four terms in Bayes' Theorem. We referred to $p[\theta|y]$ and $p[\theta]$ as 'probability distributions'. Strictly speaking, that is incorrect because as written the $p[\theta|y]$ and $p[\theta]$ are probability densities for specific values of θ, not probability distributions over its whole domain. However, Bayes' Theorem is valid for any choice of θ and y, so distributions are implied. (There is formal notation that makes distributional assumptions explicit, such as '$p_{\Theta|Y}[\theta|y]$ for $\theta \in \Theta$ and $y \in Y$', but we use the shorter form for readability.)

We can also write Bayes' Theorem with just three terms. In most applications, the value of the data y is given, and we cannot change that. But we can consider many different possible values of the parameter θ. So $p[y]$ is constant but the other three terms vary depending on the value of θ. In terms of the varying quantities, we can thus say that the posterior is proportional to the product of the prior and the likelihood:

$$\underline{\text{BAYES' THEOREM}} \text{ (proportionality form):}$$
$$p[\theta|y] \propto p[\theta]p[y|\theta]. \tag{2.3}$$

Let's see whether we can rewrite our murder mystery in this more formal language. The story is a problem of parameter estimation. There is only one parameter, θ, with just three possible values: $\theta \in \{w, s, b\}$, where $w = wife, s = son, b = butler$. So our prior is a discrete probability distribution, and we assigned the following: $p[w] = p[s] = 0.1, p[b] = 0.8$ (using shorthand notation, $p[w]$ for $p[\theta = w]$ etc.). The first piece of evidence y is the alibi of the butler, which equates—in probabilistic language—to a likelihood of zero for $\theta = b$. If the butler were guilty, then that piece of evidence could not have been found. We write that symbolically as $p[y|b] = 0$. And because Bayes' Theorem tells us that $p[b|y] \propto p[y|b]$, we find that $p[b|y] = 0$. And so on for the rest of the story.

When our parameter has a discrete distribution, we can use the odds-form of Bayes' Theorem. The odds on an event A are the probability of it happening divided by the probability of it not happening: $O[A] = p[A]/p[\neg A] = p[A]/(1 - p[A])$. Let's derive the posterior odds for a discrete parameter to have the value $\theta = a$. We again use shorthand notation: instead of writing $p[\theta = a]$, we write $p[a]$, etc. Then:

$$\underline{\text{BAYES' THEOREM}} \text{ (odds form):}$$
$$\begin{aligned} p[a|y] &= \frac{p[a]\,p[y|a]}{p[y]} \\ p[\neg a|y] &= \frac{p[\neg a]\,p[y|\neg a]}{p[y]} \\ \Longrightarrow \frac{p[a|y]}{p[\neg a|y]} &= \frac{p[a]}{p[\neg a]}\frac{p[y|a]}{p[y|\neg a]} \\ \Longrightarrow O[a|y] &= O[a]\,\frac{p[y|a]}{p[y|\neg a]}. \end{aligned} \tag{2.4}$$

So this gives us a very simple way of expressing Bayes' Theorem: *posterior odds are prior odds times the likelihood ratio*. It is especially useful for binary problems, where the parameter can only assume two values, e.g. true or false, 0 or 1, A or B, wife or son. But we can generalise it to any number of parameter values: *posterior relative probabilities are prior relative probabilities times the relative likelihoods.* That is what we used informally in the murder mystery, where for wife:son:butler our relative probabilities went from 10:10:80 via 50:50:0 to 33:67:0 (or equivalently from 1:1:8 via 1:1:0 to 1:2:0).

2.3.1 Medical Example

Let's use the odds-form of Bayes' Theorem in a simple application. It is the perhaps most popular example used in Bayesian tutorials: how to interpret a medical diagnosis. Say that an epidemic has started and 1% of people is infected. A diagnostic test exists that gives the correct result in 99% of cases. We undergo the test and the result is positive, which is bad news because for diagnosticians 'positive' means 'infected'. But what is the probability that you actually do have the disease?

Before reading on, try to answer the question yourself!

Let's proceed to the answer. As for every Bayesian analysis, we must specify our prior and likelihood. Our prior probability of being infected is 1% so our prior odds are $O[infected] = 1/99$. Our likelihood function is: $\{\ p[positive|infected] = 99\%,\ p[positive|not\ infected] = 1\%\ \}$, so our likelihood ratio is 99. Your posterior odds for having the disease are thus $O[infected|positive] = 1/99 \times 99 = 1$. These even odds mean that, perhaps surprisingly giving the high accuracy of the diagnostic test, your posterior probability of having the disease is still only 50%. This is because the small error probability of the test is matched by an equally small prior probability of having the disease.

2.4 Bayes and Science

In all the sciences, our aims are to explain observations and predict the future: see the discussion of the 'Science Equation' (Eq. (1.2) in Chap. 1). For example, in environmental science and ecology, we use predictions to assess the likely impact of environmental change on ecosystems and to optimise management. The problems are very wide-ranging in data availability and model choice. Bayes can help us in all cases to process the information optimally. Irrespective of model structure, from the simplest linear regression to the most complex computer model, Bayesian methods can be used to quantify and reduce the modelling uncertainties. The two most basic uses of Bayesian methods are to tell us which parameter values and which models are plausible given data y:

1. Parameter estimation $p[\theta|y]$, where θ is a scalar or a vector in parameter space.
2. Model evaluation $p[f|y]$, where f is a model in the set of possible models.

Every science question can be put in the form: 'What is $p[\theta|y]$?' or 'What is $p[f|y]$?' That is just the scientific way of asking the question: 'What can observations teach us about the world?' So everything is about conditional probability distributions. In a beautiful analysis, R.T. Cox proved already in 1946 that consistent reasoning, as commonly understood, implies the laws of probability theory (Jaynes, 2003). The world of Sherlock Holmes, with proofs rather than probabilities, and deduction rather than induction, is not science. Nothing can be unconditionally proven in science, there are always assumptions regarding the quality of our data and models that force us to use probabilistic language. In the words of Lindley (1991), 'Probability, it has been said, is merely common sense reduced to calculation. It is the basic tool for appreciating uncertainty, and uncertainty cannot be adequately handled without a knowledge of probability'. Jaynes (2003) put it more succinctly: 'Probability theory is the logic of science'.

Note that in our murder mystery, we allowed for the possibility that you, the detective and your assistant assigned different prior probabilities of guilt to the three suspects. This subjective aspect of probability in fact applies to all prior probabilities and likelihood functions that we assign informally. Different people have different expertise and background information, so we can expect them to assign different probabilities to events in the real world. But once we have specified our prior and likelihood, we have no choice but to accept the posterior distribution that Bayes' Theorem then gives us: there is no subjectivity in that step. In a scientific context, we need to account for all inevitable subjectivities by being transparent about them (i.e. mention which prior and likelihood we used), and where possible minimise the subjectivity by using formal methods of specifying the prior distribution. We shall come back to that issue in Chap. 3.

2.5 Bayesian Parameter Estimation

We always have some uncertainty about the proper values of a model's parameters, and we express this uncertainty as the probability distribution $p[\theta]$. When new data y arrive, we want to use those data to reduce our uncertainty about θ. Bayes' Theorem tells us that our posterior distribution $p[\theta|y]$ is proportional to the prior $p[\theta]$ and to the likelihood $L[\theta] = p[y|\theta]$. Parameter estimation using this theorem is called *Bayesian calibration* (Kennedy & O'Hagan, 2001; Van Oijen et al., 2005). At this point, we briefly remind the reader that both the parameter vector θ and the data vector y can be (highly) multi-dimensional. The posterior distribution for the parameters, $p[\theta|y]$, is then a *joint* probability distribution, and it is likely that most parameter-pairs in θ will show some degree of correlation. Let us assume, for now, that model error ϵ_{model} is much smaller than measurement error and can be ignored. We can then use the Science Equation (1.3) to derive the likelihood function:

$$\begin{aligned}&\underline{\text{LIKELIHOOD FUNCTION}}\ (\text{given } \epsilon_{\text{model}} = 0):\\ &p[y|\theta] = p[f(x,\theta) - y = \epsilon_y].\end{aligned} \tag{2.5}$$

If we would further assume that ϵ_y has a zero-mean normal distribution and a coefficient of variation of 50% (we used the same example before), then the likelihood function would simplify to $p[y|\theta] = N[f(x, \theta) - y; \mu = 0, \sigma = y/2]$. Say that we have also defined the prior distribution for the parameters, $p[\theta]$, based on literature review and expert opinion. Then the final step, as prescribed by Bayes' Theorem, is to find the product of the prior and the likelihood. In principle, a representative sample from the posterior distribution can easily be generated by Monte Carlo sampling: take a large sample from the prior and use the likelihoods as weights in deciding which parameter values to keep. More sophisticated methods such as MCMC will be explained in later chapters. This simple example captures all the essential steps in Bayesian calibration: (1) specify a prior, (2) specify $p[\epsilon_y]$ and from it the likelihood function, (3) apply Bayes' Theorem.

After a Bayesian analysis, we will have reduced our uncertainty about models and parameters. But we will of course primarily be interested in what we have learned about the real world itself. Such applied problems include forecasting, risk analysis and decision-making. In each of those cases, we are interested in $p[z]$, which we estimate as $p[f(x, \theta) + \epsilon_{model}|y]$. The main paradigm shift that Bayesian thinking brings to science is that we no longer aim to find the single best parameter value or the single best model, instead we think in terms of probability distributions.

Bayes' Theorem as written in (2.2) or (2.3) has some immediate consequences for parameter estimation. First, if you assign a prior probability of zero to certain values of the parameters θ, then whatever evidence may subsequently appear, those parameter values will always remain impossible. The product of zero and any other value is always zero. For example, if you have parameters that a priori are considered to be positive, then even a plethora of negative measurement values will not lead to negative estimates. Likewise, once you have found evidence that implies zero likelihood for certain parameter values, then those values remain excluded for ever. So we should only rule out parameter values—in their prior or likelihood function—if we are 100% certain that those values are impossible. On the other hand, if we do have that certainty, then a restricted prior can make subsequent analysis much more efficient.

Our focus in the next few chapters will be on parameter estimation ($p[\theta|y]$). We first discuss how to assign the prior and likelihood (Chaps. 3 and 4) and then how to derive the posterior (Chap. 5). In later chapters, we shall consider model comparison and selection ($p[f|y]$).

Exercises

1. **Simpson's Paradox (part 1).** You are studying the popularity of running in a group of 20 people (ten young, ten old). You wonder whether age affects the desire to run. But you notice that half your study group are smokers, and you want to control for that factor. So you begin by splitting up your group into smokers and non-smokers. You find that in both subgroups, a higher percentage

of old people go running than of young people. Does that imply that running is more popular amongst the old than the young overall? In other words, must there be more old runners than young runners in your study group of 20 people? (This exercise and the next aim to train your intuition about conditional probabilities. A more advanced treatment of this problem will be deferred until Chap. 16.)

2. **Simpson's Paradox (part 2).** You are still studying your group of 20 people, and here are the detailed results:

SMOKERS			NON-SMOKERS		
Old	Run	n	Old	Run	n
0	0	2	0	0	5
0	1	0	0	1	3
1	0	7	1	0	1
1	1	1	1	1	1

Assume that frequencies indicate probabilities. Now calculate $p[run|young]$ and $p[run|old]$ for: **a.** smokers, **b.** non-smokers, **c.** the whole group of 20 people (and explain in words the paradoxical result concerning age impact on running, when comparing the first two answers with the last one).

3. **Derivation of Bayes' Theorem.** Two basic rules of probability are the *Product Rule* and the *Law of Total Probability*. Both can be found in Appendix C. From which of the two does Bayes' Theorem follow immediately?

4. **Medical problem, different prior odds.** In the medical example, we found a posterior disease probability for you of 50% after a positive test result. What would it have been if not 1% but 5% of the population had been infected?

5. **Medical problem, different likelihood ratio.** What would your posterior disease probability have been if you had received a negative test result?

6. **Inference when data become overwhelming.** In the murder mystery, your assistant started with a different prior distribution for the murderer than you did. He suspected the wife the most. Say his prior relative probability distribution for wife:son:butler was 60:20:20 (or 3:1:1). Assuming he interprets the new information (butler alibi, murder weapon) in the same way as you do, what will be his posterior distribution? If you would keep finding new evidence, would your assistant and you eventually end up with the same 'final' posterior distribution?

Chapter 3
Assigning a Prior Distribution

As we saw in the murder mystery of the preceding chapter, there is no such thing as *the* prior probability distribution of a parameter or a set of models. A prior expresses uncertainty arising from incomplete knowledge, and whatever the subject is, people have different knowledge and expertise. So instead of speaking of 'the prior probability of x', each of us should say 'my prior probability for x'. We *assign* a prior probability distribution; we do not *identify* it. This is even the case when we invite the opinion of experts on the likely values of our model's parameters. There is an art and science to eliciting expert opinion in order to formulate a Bayesian prior (Mikkola et al., 2023; O'Hagan, 2012), but ultimately the responsibility for the prior lies with the modeller, not the expert panel.

We do not have complete freedom in assigning just any distribution. First of all, we are working within probability theory, so we must make *coherent* probability assignments (Jaynes, 2003; Lindley, 1991). If we say that $p[A] = 0.5$, then we cannot say that $p[A \text{ or } B] = 0.4$. In general, if we have multiple parameters in our model, then our task a priori is to assign a valid *joint* probability distribution to all parameters.

Moreover, when you and I have exactly the same information about, say, a parameter θ, then we should be assigning the same probability distribution $p[\theta]$. When the parameter is by definition non-negative, as most physical properties are, then a probability distribution such as the Gaussian which ranges from minus to plus infinity is inappropriate. However, we need not be too principled here. A Gaussian distribution $N[\mu, \sigma^2]$ with positive mean μ and a typical coefficient of variation σ/μ of 20%, will have less than one-millionth of its probability mass below zero. Also, when you are expecting many data to be forthcoming, then the role of the prior becomes small: the posterior distribution will then be mainly determined by the likelihood function. However, there is a tendency among researchers to select too wide, uninformative prior distributions. An extreme example of this is *maximum likelihood estimation* (MLE) which does not require you to specify any

M. van Oijen, *Bayesian Compendium*,
https://doi.org/10.1007/978-3-031-66085-6_3

prior distribution at all, and behind the scenes is assuming that any parameter value is a priori equally plausible. This is bad use of information and should generally be avoided. If you have enough information about a system to build a model for it, with parameters that you have given implicit meaning by the role they play in your model, then it is not plausible to have no idea about parameter values at all. As a simple example, if you believe that $f(x, \theta) = x\theta$ is a proper model for your system, then you are saying that output $f()$ responds linearly to x. Could you make such a statement without having any idea about whether the proportionality is negative, zero or positive?

3.1 Quantifying Uncertainty and MaxEnt

To avoid the opposite extremes of ignoring parameter uncertainty and assuming it infinite, the general principle in assigning priors should be to choose that distribution which expresses maximum uncertainty while being constrained by whatever information you have. But that raises the question: what is the appropriate quantitative measure of uncertainty? We can immediately rule out the *variance*, as a simple example will show. Take a fair die, such that our uncertainty about the number of dots showing after the next throw is uniformly distributed as $p[i] = \frac{1}{6}; i \in \{1..6\}$. That complete uncertainty has a variance $\text{Var}[i] = 2.92$, whereas a much less uncertain distribution with $p[1] = p[6] = 0.5$ and $p[i \in \{2, 3, 4, 5\}] = 0$ has the maximum possible variance for a six-sided die of 6.25. A much better measure of uncertainty is the *entropy* S of the distribution, defined for discrete probability distributions as:

ENTROPY OF DISCRETE DISTRIBUTION:

$$S = -\sum_{i=1}^{n} p[i] \ln(p[i]), \tag{3.1}$$

and a similar equation exists for continuous distributions with an integral replacing the summation. Just like Cox showed formally that consistent rational reasoning requires us to follow the rules of probability theory, Jaynes (2003) showed that for uncertainty we should use entropy. His *Maximum Entropy* (MaxEnt) principle for assigning probability distributions has become a standard approach in many areas of science, and it has even been implemented in an R-package called FD by Laliberté and Shipley. Let's try it out on the example of the fair die that we have just given. We ask for the discrete probability distribution over $\{1..6\}$ with maximum entropy under the constraint that the mean is equal to 3.5. That can be answered with the following very short R-code:

```
Dp  <- 1:6 ; meA <- FD::maxent(3.5, Dp)
pA <- meA$prob ; SA <- meA$entropy
```

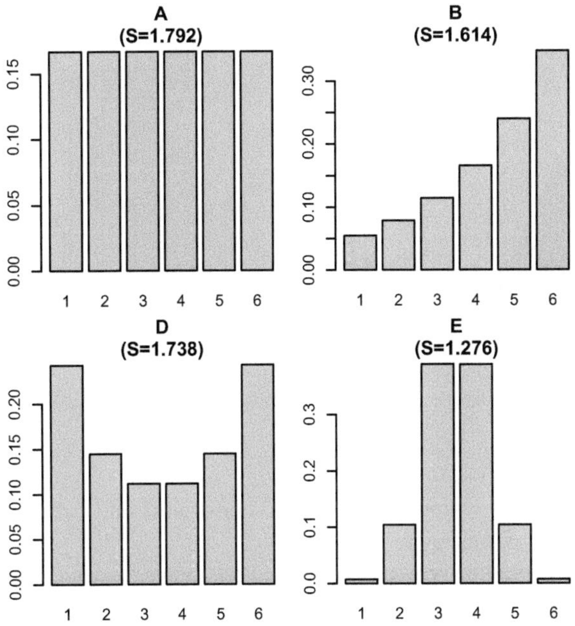

Fig. 3.1 Distributions selected using MaxEnt. Entropies between brackets. All distributions are on the same domain (1,..,6) but with different constraints for mean and mean square: see text

This code gives `pA` as the MaxEnt distribution, which we plot in panel A of Fig. 3.1. It is, as we expected, the uniform distribution which has entropy equal to `SA` = 1.792.

The other panels in Fig. 3.1 show the MaxEnt distributions for when we have different information. Panel B assumes a mean of 4.5 which makes the MaxEnt distribution slope upward. Panel C assumes a mean value for i^2 of 13. Panel D assumes both a mean for i of 3.5 and for i^2 of 16. Panel E is as D but with mean square 13 instead of 16.

We show two more MaxEnt distributions in Fig. 3.2. The panel marked F has the same constraints as panel E in the first figure, but we have refined the domain from $\{1, .., 6\}$ to $\{1.0, 1.1, .., 5.9, 6.0\}$. So we have gone from a six-sided die to a 51-sided one. You see that the distribution then becomes far more bell-shaped. In the limit of a continuous distribution with prescribed mean and mean square (or variance), MaxEnt identifies a Gaussian distribution as the most uncertain one. So if all you learn about a parameter is the first two *moments* of its distribution, MaxEnt suggests that you assign a Gaussian distribution. Note further that entropies cannot be compared for distributions over different domains: entropy values S are higher for distribution F than for E only because of its greater domain multiplicity.

The final panel G shows an elegant result for a 10-sided die. The depicted distribution was identified by MaxEnt given constraints for the mean of i and the mean of i^3 (third moment), and this produced the beautiful wave pattern as the most uncertain distribution. I leave you to verify for yourself that the distributions in the

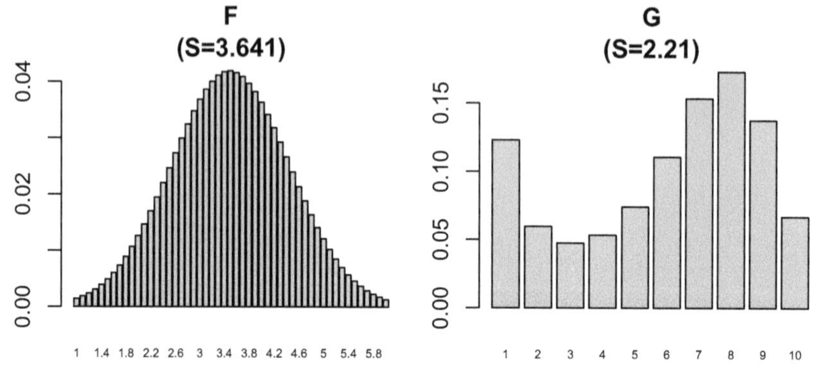

Fig. 3.2 More MaxEnt distributions. Domains cover 51 and 10 possible values. Entropies between brackets

two figures have maximum entropy within their different constraints and perhaps explore some more distributions of your own.

There is a rich literature on maximum entropy and related principles for assigning prior probability distributions, and Jaynes (2003) is a very good place to start, as is the older work of Jeffreys (see e.g. Robert & Rousseau, 2009). An important special class of parameters identified by Jeffreys is that of *scaling* parameters, such as standard deviations σ and variances σ^2. These are non-negative parameters that indicate the scale over which other variables vary, and Jeffreys and Jaynes showed that, without other knowledge, the prior density for such parameters should decrease inversely with their magnitude, i.e. $p[\sigma] \propto \frac{1}{\sigma}$ defined on the whole domain from zero to plus infinity. This is an example of a so-called *Jeffreys Prior*. However, we often do have more information about the variance than just non-negativity, and then, the Jeffreys Prior will not be the right choice.

3.2 Priors in Practice

Bayes is ubiquitous: the same Bayesian methodology is used throughout the sciences. However, practical problems can vary between different types of application. In disciplines where prior information is scarce, say in the study of rare species, we can only assign wide, uninformative prior distributions, and we need to be very careful about what the data tell us about the relative likelihoods of different parameter values. In other disciplines, say agriculture, there may be rich prior information but from disparate sources, making it difficult to come up with a proper prior distribution. It may also be the case that you are modelling the behaviour of a population of organisms or variables that are dispersed over different locations. In such cases, a *hierarchical prior* may be useful as we shall discuss in Chap. 17.

Whatever the circumstances, in most cases there will be a degree of apparent subjectivity in the choice of prior, because people have different background knowledge. Therefore, in publications, it is important to state clearly what prior you have chosen and what information you based it on. It will also be a service to the reader if you show how much the results of your Bayesian calibration change when choosing a different prior.

The next chapter considers the second element in Bayes' Theorem: the likelihood function.

Exercise

1. **Family matters (prior).** You are a researcher of family size. You assign a Jeffreys prior to the number of children N in a family, but you assume that N is 6 at most. Then you meet Tom. **a.** What are your prior odds for $N \in \{1..6\}$ in Tom's case? Express the odds in integers. **b.** What is your prior probability that Tom is an only child? **c.** And what is the probability that he is an eldest child?

Chapter 4
Assigning a Likelihood Function

As scientists, we want to parameterise our models and compare them against alternative models. For these purposes, measurement data are needed, but how exactly should we use the data? The answer is always the same: in the *likelihood function*. The likelihood is a conditional probability density function, denoted as $p[y|\theta]$ (or more succinctly as $L[\theta]$), where as usual y can be multi-dimensional. It is the answer to the question: 'what is the probability of measuring y if the true value is $f(x,\theta) + \epsilon_{\text{model}}$?'. This can be written formally as follows:

$$\underline{\text{LIKELIHOOD FUNCTION:}} \quad L[\theta] = p[y|\theta] = p[f(x,\theta) - y = \epsilon_y + \epsilon_{\text{model}}]. \tag{4.1}$$

This definition of the likelihood function follows from the Science Equation (Eq. (1.3)), and in Chap. 2 we already gave a simplified definition of the likelihood function where ϵ_{model} was assumed to be zero. We shall largely ignore model structural error in this chapter as well but return to that issue in Chaps. 11 and 12.

We use the likelihood function, together with the prior distribution discussed in the preceding chapter, in Bayes' Theorem to derive the posterior distribution for our parameters. But, just as with the prior distribution, we do not *derive* the likelihood function, we have to *define* it ourselves. So there is an element of subjectivity here: our choice of likelihood function will depend on what we believe the data can tell us. And that will depend on our views of the possible errors in the data. I would say that in practice, the hardest and most important step in Bayesian calibration is formulating an appropriate likelihood function.

The likelihood is a powerful concept: y is usually multi-dimensional and may consist of measurements on several different variables at various times and locations, yet $p[y|\theta]$ can always be defined. All we need is a model that links the parameters to predictions. It is through the likelihood function that Bayesian calibration has the capacity to use highly heterogeneous datasets in parameter estimation. Levy et al. (2017) used Bayesian calibration to reconcile

M. van Oijen, *Bayesian Compendium*,
https://doi.org/10.1007/978-3-031-66085-6_4

eddy-covariance measurements of N_2O emissions with chamber measurements of the same fluxes. Patenaude et al. (2008) combined remote sensing data from satellites with field-based data on Corsican pine stands in the United Kingdom, in Bayesian calibration for the parameters of forest PBM 3-PG. Höglind et al. (2016) combined measurements on 10 different variables from five grassland sites in their likelihood function. With such rich datasets, it becomes important to assess whether measurement errors for different variables have correlations that should be expressed in the likelihood function.

Formulating the likelihood function can be difficult, even when model error ϵ_{model} can be ignored and different variables are measured independently, because measurements can be wrong in three different ways, as summarised in the next equation:

$$\epsilon_y = \epsilon_{y,\text{stochastic}} + \epsilon_{y,\text{systematic}} + \epsilon_{y,\text{representativeness}}. \tag{4.2}$$

The first of the three data error terms, $\epsilon_{y,\text{stochastic}}$, quantifies random measurement noise, independent for each data point in y, which we could represent with a zero-mean normal distribution (Ogle & Barber, 2008). This is often the only data error recognised by modellers, but it is not always the most important one. The second error term, $\epsilon_{y,\text{systematic}}$ represents measurement bias which could shift whole collections of data up- or downward (beautifully illustrated by Jaynes (2003, pp. 257–260) with his emperor of China parable). The final term, $\epsilon_{y,\text{representativeness}}$, is generally the hardest to quantify. It refers to the possibility that our data derive from other conditions than our model is designed for. Take the example of a model for crop growth. If the observed crop has a hidden growth limitation, say phosphorus deficiency, that is not expressed in the model, then the data—from the point of view of model parameter estimation—will be underestimating growth. In contrast to the other two types of measurement error, the scope for $\epsilon_{y,\text{representativeness}}$ cannot be reduced by more careful measurement or greater sampling intensity.

For every dataset, the modeller and data expert need to distinguish the three types of possible data error and assign probability distributions to the respective error uncertainties. The parameters of these probability distributions, such as the standard deviation of the stochastic noise, can be fully specified a priori (e.g. $\sigma_{y,\text{stochastic}} = y/2$), or they can be included with the other parameters to be estimated in the calibration. The latter method would be preferable in this example if we have little information about the precision of our measurement instrument. Van Oijen et al. (2011) calibrated the degree of systematic error in chamber measurements of soil-fluxes of CO_2, N_2O and NO using four different forest PBMs, and all four calibrations suggested that the measured CO_2 emission values had been unrealistically high.

4.1 Expressing Knowledge About Data Error in the Likelihood

When calibrating their model, many modellers as mentioned assume that their model is correct and that their uncertainty about every data point can be described by a Gaussian distribution with constant variance σ_y^2. Their likelihood function is then:

<u>GAUSSIAN LIKELIHOOD FUNCTION</u> for independent errors:

$$p[y|x, \theta, f] = |2\pi \Sigma_y|^{-1/2} \exp\left(-\frac{1}{2}(f(x, \theta) - y)^\top \Sigma_y^{-1}(f(x, \theta) - y)\right)$$

$$= \prod_{i=1}^{n_y} N[f(x_i, \theta) - y_i; 0, \sigma_y^2], \tag{4.3}$$

where the covariance matrix has σ_y^2 everywhere on the diagonal and zeros elsewhere: $\Sigma_y = \sigma_y^2 \mathbf{I}_{n_y}$. That commonly used Gaussian function presumes that all measurement errors are independent from each other, and that measurement uncertainty is the same irrespective of the value of y. But those assumptions generally do not do justice to everything we know. We often have information about the measurement process that we can include in the covariance matrix Σ_y. One important example is that spatially or temporally nearby observations may have similar measurement error. Our measurement instrument may have drift and conditions for measurement may be harder in certain areas than in others. We can express such error correlations in the off-diagonal terms of Σ_y (Gregory, 2005; Sivia & Skilling, 2006) using ideas from Gaussian Process modelling (Chap. 15) and time series analysis (Chap. 22).

It will usually also be better not to have the exact same value everywhere on the diagonal of Σ_y because measurement error is likely to vary with the magnitude of the y_i. Estimating the weight of an elephant comes with greater uncertainty variance than estimating the weight of a mouse. So our uncertainty about y increases with its magnitude, but how strongly? The opposite extreme to setting the variance σ_y^2 as a constant is assuming a constant coefficient of variation ($c.v. = \sigma/y$), which is equivalent to the variance being proportional to y^2. That choice may underestimate our uncertainty for small y. A more reasonable intermediate position is to have our variance be proportional to y.

Let's compare these three approaches in a concrete example. We have data (x, y) with $x = (1, 2, 3)$ and $y = (10, 1, 5)$, and want to fit a straight line. Let's assume that we have no prior knowledge at all about the intercept and slope—which is of course unrealistic but serves to highlight the role of the likelihood in the calibration. Let's define the three Gaussian likelihoods as above, with always a variance of 1 for the first measurement point: $\Sigma_y[1, 1] = 1$. For the two other points we have our different choices:

1. variance independent of y: $\Sigma_y[2, 2] = \Sigma_y[3, 3] = 1$,
2. variance proportional to y: $\Sigma_y[2, 2] = 0.1$; $\Sigma_y[3, 3] = 0.5$,
3. variance proportional to y^2: $\Sigma_y[2, 2] = 0.01$; $\Sigma_y[3, 3] = 0.25$.

We do the Bayesian calibration for these three choices and that gives us the following posterior distributions $N[\mu_{\beta|y}, \Sigma_{\beta|y}]$ for the parameter vector $\beta = (intercept, slope)$. (We leave the mechanics of how to find that posterior to the following chapters.)

<u>CALIBRATION WITH 3 LIKELIHOOD FUNCTIONS:</u>

$$\begin{aligned}
\text{Case 1: } \mu_{\beta|y} &= \begin{bmatrix} 10.33 \\ -2.5 \end{bmatrix}; \quad \Sigma_{\beta|y} = \begin{bmatrix} 2.33 & -1 \\ -1 & 0.5 \end{bmatrix}. \\
\text{Case 2: } \mu_{\beta|y} &= \begin{bmatrix} 3.95 \\ -0.79 \end{bmatrix}; \quad \Sigma_{\beta|y} = \begin{bmatrix} 1.55 & -0.71 \\ -0.71 & 0.34 \end{bmatrix}. \\
\text{Case 3: } \mu_{\beta|y} &= \begin{bmatrix} -1.36 \\ 1.28 \end{bmatrix}; \quad \Sigma_{\beta|y} = \begin{bmatrix} 0.85 & -0.41 \\ -0.41 & 0.2 \end{bmatrix}.
\end{aligned} \tag{4.4}$$

So we see very large differences between the distributions! In Fig. 4.1, we show the three lines that are represented by the posterior mean parameter vectors. The figure shows that with likelihood function 1, the first and highest data point is very influential, whereas that point is largely ignored in calibration 3. Case 2 may be the happy intermediate, but it depends of course on your knowledge about the measurement process.

Note that you can always decide to estimate the σ_y^2 in your calibration rather than prescribe them as we did here. That approach may be a wise choice if you have little prior knowledge about the variability of measurement error. But you then obviously need to assign a joint prior probability distribution to the σ_y^2 and should still try to include in that prior whatever you can say about the measurement error you expect.

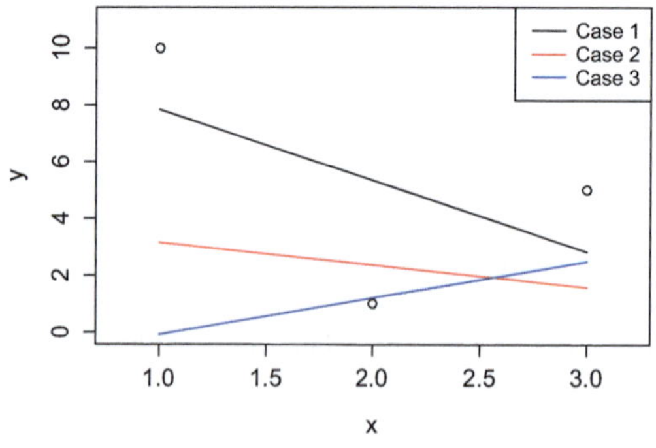

Fig. 4.1 Posterior mean straight lines derived by Bayesian calibration using three different likelihood functions. Measurement variance is constant (Case 1), proportional to y (Case 2), proportional to y^2 (Case 3)

We already mentioned that we can represent correlated uncertainties about data points in the off-diagonal terms of the covariance matrix. That is a specific example of a more general important consideration when formulating our likelihood function: we should not exclusively focus on the values of individual data points but evaluate *patterns* of system behaviour. For example, if the values of y show an oscillating pattern over time, then parameter vectors that instead lead to stabilising behaviour should be accorded low likelihood. Spectral analysis of y and analysis of statistical moments may help us quantify patterns succinctly.

4.2 What to Measure

This chapter has discussed the likelihood function as the means by which empirical information enters into our analysis. But we have not discussed what data to gather in the first place as that is for the most part too problem-specific an issue. But we can use the Bayesian approach to evaluate different measurement strategies (Bon et al., 2023). For example, when considering replacing a piece of equipment with a higher-precision tool that is more expensive, we could first run Bayesian calibrations with virtual data of differing precision in order to assess how much the new tool is likely to reduce predictive uncertainty (Van Oijen et al., 2005). And during the course of your experiments and monitoring studies, you may want to change your measurement set-up if it no longer gives you the information that you need. That is the topic of Bayesian adaptive sampling which will be discussed in Chap. 24. More generally, the concept of 'Value of Information' is central to Bayesian decision theory (BDT, see Chap. 19).

Exercises

1. **Family matters (likelihood).** You are a researcher of family size. You meet Tom and he tells you that he is the n-th eldest child in his family. What is the likelihood function for the total number of children N in his family?
2. **Correlated data errors.** Say you are carrying out linear regression of y on x. The measurement errors for y are strongly positively correlated. But you don't know this, so you have not specified any correlations (non-zero covariances) in the measurement uncertainty matrix Σ_y that appears in your Gaussian likelihood function (you are using Eq. (4.3)). The error correlations will of course affect the quality of your parameter estimation. Do you expect the intercept and slope to be estimated better or worse than with independent measurement errors? The answer may surprise you!

Chapter 5
Deriving the Posterior Distribution

In the preceding chapters, we discussed how we can assign a prior distribution for our parameters and how to choose a likelihood function that captures the information content of our data. So all that is left is to apply Bayes' Theorem (Eq. (2.2)) to derive our desired posterior distribution. Note that when talking about the posterior, we use the phrase 'deriving the' distribution rather than 'assigning a' distribution. That is because Bayes' Theorem tells us exactly what the posterior distribution should be once we have defined our prior and likelihood.

Even though Bayes' Theorem is simple to write down, the right-hand side is not in a convenient form: it is often not easy to see what kind of probability distribution is defined by the product of prior and likelihood. So how do we use Bayes' Theorem in practice? Well, there are two ways of using the theorem to find the posterior distribution: *analytically* and *numerically*. The analytical method is the truly mathematical way, it consists of writing down the formula for the posterior distribution directly. But this can only be done in certain simple cases, e.g. when both the prior and the likelihood are Gaussian distributions. The posterior distribution is then also a Gaussian. There are a few other combinations of probability distributions that allow you to directly write down the posterior distribution (which then always belongs to the same type as the prior), but we shall not study many of such so-called 'conjugate' distribution pairs in this book. They are generally not good reflections of the information that we have a priori or receive in new data. However, we should be aware of these methods because they are implicitly used by standard linear models, so we do work out an example in the following section.

M. van Oijen, *Bayesian Compendium*,
https://doi.org/10.1007/978-3-031-66085-6_5

5.1 Analytically Solving Bayes' Theorem: Conjugacy

Bayes' Theorem says that the posterior is proportional to prior times likelihood. As mentioned, we can calculate the posterior analytically if the prior and likelihood belong to conjugate distributions. A simple example is when we want to estimate a scalar parameter for which both prior and likelihood are univariate Gaussians. Then the posterior will also be a univariate Gaussian. Say we have the following definitions for prior $p[\theta]$ and likelihood $p[y|\theta]$ where the dataset y consists of n measurements y_i of the parameter, each with the same measurement uncertainty σ_y^2:

$$p[\theta] = \frac{1}{\sigma_0\sqrt{2\pi}} \exp\left(-\frac{(\theta-\mu_0)^2}{2\sigma_0^2}\right), \tag{5.1}$$

$$p[y|\theta] = \prod_{i=1}^{n} \frac{1}{\sigma_y\sqrt{2\pi}} \exp\left(-\frac{(y_i-\theta)^2}{2\sigma_y^2}\right). \tag{5.2}$$

In that simple case, the posterior distribution $p[\theta|y]$ is:

$$p[\theta|y] = \frac{1}{\sigma_1\sqrt{2\pi}} \exp\left(-\frac{(\theta-\mu_1)^2}{2\sigma_1^2}\right), \tag{5.3}$$

where

$$\begin{aligned}
\mu_1 &= \frac{1/\sigma_0^2}{1/\sigma_0^2 + n/\sigma_y^2}\mu_0 + \frac{n/\sigma_y^2}{1/\sigma_0^2 + n/\sigma_y^2}\bar{y}, \\
\frac{1}{\sigma_1^2} &= \frac{1}{\sigma_0^2} + \frac{n}{\sigma_y^2}, \\
\bar{y} &= \frac{1}{n}\sum_i y_i .
\end{aligned}$$

Let's make this example concrete by assuming some values for the data and the prior. We choose data y = (1.0, 0.9, 1.1) with variance Vy = 1, and for the prior we choose a mean m0 = 0 and a variance V0 = 1. This is the R-code for these choices that then calculates the posterior:

```
m0 <- 0               ; V0 <- 1
y  <- c(1.0,0.9,1.1)  ; ny <- length(y)
my <- mean(y)         ; Vy <- 1
k  <- (1/V0) / (1/V0 + ny*Vy)
m1 <- k*m0 + (1-k)*my ; V1 <- 1 / (1/V0 + ny/Vy)
```

When we run this code, we find that the posterior mean `m1` is 0.75, and the posterior variance `V1` is 0.25. You can verify that that is consistent with the formulas shown earlier. In the next section, we show how to estimate the posterior for this problem numerically.

Later in this book, we will encounter two other examples of conjugacy: a beta prior with a binomial likelihood (Chap. 18) and a generalisation thereof, a Dirichlet prior with a multinomial likelihood (Chap. 21).

5.2 Numerically Solving Bayes' Theorem: Sampling

When we cannot find the posterior analytically, we have to use numerical methods, and that is the usual case. In the example of this chapter, an analytical solution was of course possible, but let's tackle the problem numerically anyway. The numerical approaches always involve *sampling*. They do not give us a nice formula for the posterior distribution but a representative sample from it. There are many different algorithms for numerically sampling from a posterior distribution, but in this chapter we are going to use the simplest one, which is called the *Accept-Reject* (A-R) algorithm or *rejection sampling*. The A-R algorithm begins by taking a large sample from the prior which it then thins out to produce a representative sample from the posterior. It works as follows:

1. Calculate the largest possible likelihood value L_{max}.
2. Take a large sample $\{\theta_i\ ;\ i = 1, .., n\}$ from the prior distribution for the parameters.
3. For each parameter vector θ_i, draw a random number from the standard uniform distribution $u_i \sim U[0, 1]$ and accept θ_i into the final sample if $u < L[\theta_i]/L_{max}$.

Let's implement this in R, using $n = 10^5$, and do the sampling:

```
Lmax       <- prod( dnorm( y, my, sqrt(Vy) ) )
n          <- 1e5 ; samplePrior <- rnorm( n, m0, sqrt(V0) )
L          <- function(b){ prod( dnorm( y, b, sqrt(Vy) ) ) }
Lrelative  <- sapply(samplePrior,L) / Lmax
iAccept    <- which( runif(n) < Lrelative )
samplePost <- samplePrior[ iAccept ]
m1.n       <- mean( samplePost ) ; V1.n <- var( samplePost )
```

The results of the A-R sampling are shown in Fig. 5.1. We see that we get nearly the same results as the analytical solution, but there is some slight sampling error. The A-R algorithm is not very good and should only be used for very simple, low-dimensional problems. Imagine sampling from a 20-dimensional parameter space with the A-R algorithm. To sample each 'corner' of that space just once would already require a sample size of $n = 2^{20} \approx 10^6$. There are modifications of the A-R algorithm, such as *Importance Sampling*, which do not have the wasteful rejection step, but for efficiency we should really look at a different class

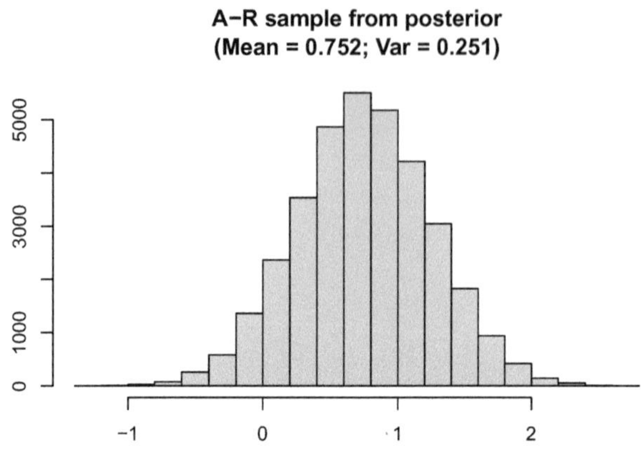

Fig. 5.1 Sampling from the posterior distribution of this chapter by means of Accept-Reject

of numerical distribution sampling methods called *Markov Chain Monte Carlo* (MCMC) methods, and that is the subject of the next two chapters.

Exercises

1. **Family matters (posterior).** You are a researcher of family size. You assign a Jeffreys prior to the number of children N in a family, but you assume that N is 6 at most. Then you meet Tom and he tells you that he is the n-th eldest child in his family. What is now your posterior probability distribution for N in Tom's family?
2. **Inference from the analytical formula.** The analytical solution of Bayes' Theorem (Eq. (5.3)) can make us quickly check certain general properties of Bayesian thinking.

   - If you set n to zero, what do you get?
   - If you take the limit of infinite data availability ($n \to \infty$), what do you get?

3. **Explore the A-R algorithm.** A downside of the A-R algorithm is that it requires a very large initial sample of size n from the prior. Rerun the A-R code with smaller values of n to see at which point sampling error becomes large (say >1%).

Chapter 6
Markov Chain Monte Carlo Sampling (MCMC)

6.1 Sampling from a Probability Distribution

The Bayesian approach to parameter estimation requires modellers to make a major mental shift: we no longer aim to find a single 'best' parameter vector—instead we aim to determine the posterior probability distribution for the parameters. Only the full probability distribution adequately represents our state of knowledge. Although this shift in thinking has made rigorous uncertainty quantification possible, it has also created computational problems. When models have many parameters, the distribution will be high-dimensional and may not be available in closed form, i.e. we cannot derive a formula for it. Often, all we can do is generate a sample from the distribution, and even that can be difficult. A solution for the problem of sampling from probability distributions that are not in closed form was provided by Metropolis et al. (1953). They introduced the so-called Markov Chain Monte Carlo (MCMC) method. MCMC is the workhorse of computational Bayesian statistics, and by now, many different MCMC algorithms exist.

MCMC consists of a walk-through parameter space in such a way that the visited points together form a representative sample from the distribution that we are after. The method of MCMC requires that at each step we decide to accept or reject a proposed new point (= vector of parameter values) depending on the relative probabilities of the current parameter vector and the proposed new one. Many different variants of MCMC now exist following this general plan. A simple introduction to the method, with an example for a process-based vegetation model, was provided by Van Oijen et al. (2005), but you will find explanations of MCMC in nearly every Bayesian text, including this chapter.

Sampling always comes with error. But MCMC algorithms are guaranteed to converge—if we keep increasing the sample size—to the true distribution. That means that the properties of our sample (mean, mode, median, variance, etc.) will get arbitrarily close to those of the true distribution, provided we keep sampling long enough. A Bayesian will mainly use MCMC to generate a representative

M. van Oijen, *Bayesian Compendium*,
https://doi.org/10.1007/978-3-031-66085-6_6

sample from the posterior distribution, but MCMC can be used to sample from any probability distribution you may be thinking of.

It is high time to show an MCMC algorithm in more detail, so let us study some code.

6.2 MCMC in Two Lines of R-Code

To study the general properties of MCMC-algorithms, we start with sampling from a *Bernoulli distribution*, which is a univariate distribution for a parameter b, where the parameter can only assume two possible values: 0 or 1. This is the simplest possible probability distribution, and you could use it to model the throw of a coin. We choose the Bernoulli distribution with even odds, which is denoted as a Br[0.5] distribution, and sample from it with the following two lines of code:

```
ni <- 10 ; b <- rep(NA,ni) ; b[1] <- 0 ; pjump <- 0.5
for(i in 2:ni){ jump <- (runif(1)<pjump) ; b[i] <- (b[i-1]+jump)%%2 }
```

We plot the progress of the sampling in the top two panels of Fig. 6.1. This is our first MCMC in action! The parameter space is extremely simple here, it consists of just the two points 0 and 1. The characteristic 'walk through parameter space' of MCMCs is shown on the left in the so-called *trace plot*. The walk consists of short sequences of one or more 0's alternating with one or more 1's. We can also see that our chain is too short: our sample from the distribution consists of seven 0's and three 1's whereas a representative sample from a Br[0.5] distribution should have equal numbers of 0's and 1's. In MCMC jargon, we say that the chain has not yet *converged*. But our chain *mixes* well, it does not get stuck in one part of parameter space. And if we run the chain for 1000 iterations, we do get nearly equal numbers of 0's and 1's, as the bottom two panels of Fig. 6.1 show.

Our MCMC mixes well because we have chosen a good *proposal distribution*, which is the probability distribution that is applied at each iteration to determine where the chain will try to go next. In our code, the proposal distribution is actually also a Br[0.5], which means that the chain will 50% of the time propose to visit the 'other' point in parameter space, and in this particular MCMC that proposal is always accepted. (That makes our MCMC an example of a so-called *Bernoulli stochastic process* that could also be produced in basic R with the command `rbinom(1000, 1, 0.5)`.)

So what do you think would happen if we had chosen a Br[0.01] proposal distribution, with only 1% jump probability at each iteration of the MCMC? The results are shown in Fig. 6.2, where you can see that the algorithm now tends to get stuck for long sequences in either the 0 or the 1 position. In other words, mixing is poor and after 1000 iterations we have clearly not yet converged. But if you would continue this poorly mixing chain long enough (say for 10^5 iterations), you would finally see near-equal proportions of 0's and 1's. So even poorly mixing MCMCs

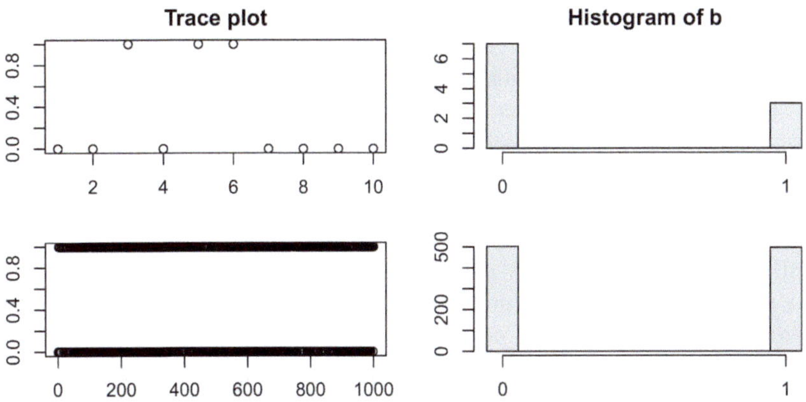

Fig. 6.1 Sampling from a Bernoulli distribution using MCMC. Top panels: MCMC with 10 iterations; bottom panels: 1000 iterations. Left: trace plots; right: chain histograms

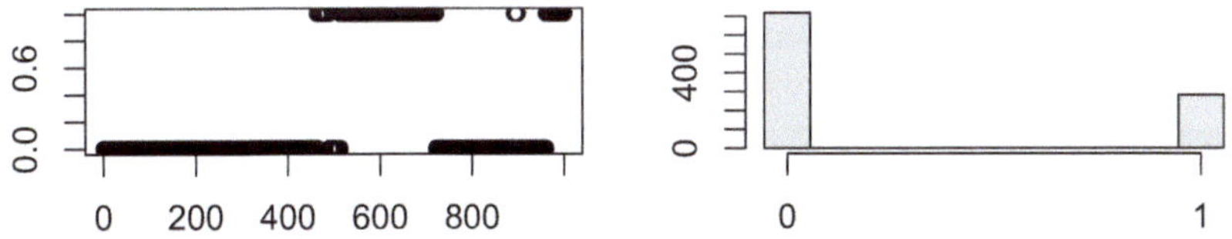

Fig. 6.2 Sampling from a Bernoulli distribution using a poorly mixing (jump probability 0.01) MCMC with 1000 iterations. Left: trace plot; right: chain histogram

are guaranteed to converge in the end, which is one of the reassuring properties of MCMC.

6.3 The Metropolis Algorithm

Now we come to a more full-fledged MCMC-algorithm which was proposed by Metropolis et al. (1953) over 70 years ago but has only found widespread application over the past three decades, with the advent of fast computers. (Other MCMC-algorithms will be discussed in Chaps. 9 and 14.) Let's jump right in and show you R-code for the Metropolis algorithm:

```
Metropolis <- function( p, b0=mb, SProp=Sb/100, ni=1e4 ){
  bChain <- matrix( NA, nrow=ni, ncol=length(b0) ) ; bChain[1,] <- b0
  for( i in 2:ni ){
     b1 <- bChain[i-1,] + as.numeric( rmvnorm(1,sigma=SProp) )
     if( runif(1) < p(b1)/p(b0) ) bChain[i,] <- b0 <- b1
     else                         bChain[i,] <- b0 }
  return(bChain) }
```

You see that the code is very short, but it is remarkably powerful, and we shall only marginally need to change it later. As you see from the '`rmvnorm`', our implementation of the algorithm uses a Gaussian proposal distribution, but you are free to use any distribution as long as it does not skip parts of parameter space by design. The Metropolis algorithm can sample from any continuous univariate or multivariate probability distribution p as long as it can calculate the probability ratio $p[b_1]/p[b_0]$ for every pair of parameter vectors b_0 and b_1 in the domain of p. If you have a formula that tells you for any b the value of $p[b]$ up to a proportionality constant, then you can of course calculate every ratio $p[b_1]/p[b_0]$ because the proportionality constant drops out. The Metropolis algorithm can be described in one sentence as 'a walk through parameter space where at each step a candidate new position $b_{\text{candidate}}$ is generated randomly and accepted with probability $p[b_{\text{candidate}}]$ / $p[b_{\text{current}}]$ (and with guaranteed acceptance if $p[b_{\text{candidate}}] > p[b_{\text{current}}]$)'. That single sentence also accounts for the meaning of 'MCMC': the 'Monte Carlo' refers to the randomness of the proposal generation, and the 'Markov' refers to the fact that the past history of the chain does not matter for its future behaviour, knowledge of the current position is always enough.

Let's make this concrete with an example. Say we want a sample from an unknown univariate probability distribution $p[\theta]$ on $[-\infty, \infty]$ where all we know is that the probability decreases exponentially with θ squared. In other words, $p[\theta] \propto \exp(-\theta^2)$. Now provide a representative sample of magnitude 10^4 from this distribution using the Metropolis algorithm. Well, we already implemented the aforementioned algorithm, so we can do this with only two more lines of R-code:

```
fA <- function(b){ exp(-b^2) }
bA <- Metropolis( fA, b0=0, SProp=diag(1) )
```

And we show the results in Fig. 6.3.

In this case, you could instead of MCMC have used basic R to sample from the distribution: `bdirect <- rnorm(1e4,0,sqrt(2)/2)`. In other words, our Metropolis sampled from a Gaussian distribution with mean zero and standard deviation equal to $\sqrt{2}/2$.

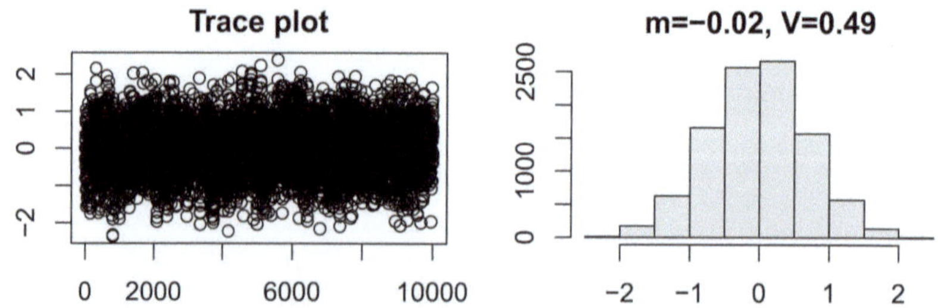

Fig. 6.3 Sampling from an unknown distribution using the Metropolis algorithm, where all we know is that p[b] is proportional to $\exp(-b^2)$

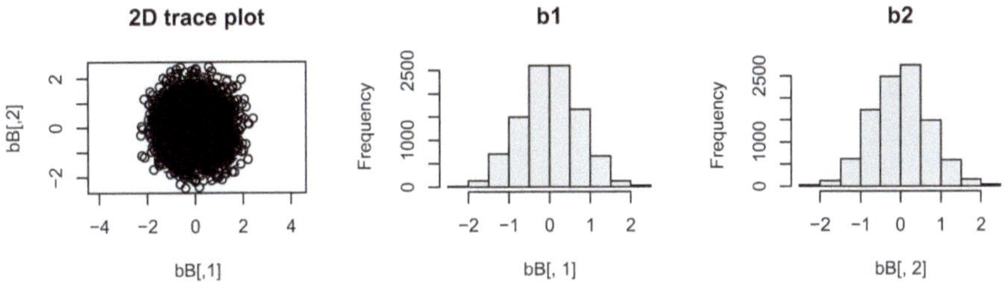

Fig. 6.4 Sampling from an unknown bivariate distribution using the Metropolis algorithm (see text)

Let's now show that the same Metropolis algorithm can also sample from unknown multivariate distributions. This time our $p[\theta]$ is a bivariate distribution on the real number plane, with $p[\theta] \propto \exp\left(-(\theta_1)^2 - (\theta_2)^2\right)$. Our code thus is:

```
fB <- function(b){ exp(-b[1]^2 - b[2]^2) }
bB <- Metropolis( fB, b0=c(0,0), SProp=diag(2)  )
```

And this produces the results of Fig. 6.4.

And also this distribution could have been sampled with an existing R-function: `rmvnorm(1e4, c(0, 0), diag(2)/2)`. Do you recognise the distribution? You need to be careful here because of the idiosyncrasies of R. The multivariate Gaussian sampling function '`rmvnorm`' requires the user to specify parameter variances and covariances—in the form of the covariance matrix—whereas the equivalent function for univariate Gaussians '`rnorm`' requires the user to specify the standard deviation, not the variance! [Even worse, later we shall encounter R-software that uses the *inverse* of the variance ...]

Our final example will show that the Metropolis algorithm will also sample correctly from multimodal distributions. This time what we know about our unknown probability distribution is the following:

- $p[\theta]$ is a bivariate, disjoint, unknown distribution on the real number plane,
- $p[\theta] = c > 0$ if the distance from (0, 0) is less than 1, and if the distance from (2, 1/2) is less than 1/2. Otherwise $p[\theta] = 0$.

We code this in the usual way and show the results in Fig. 6.5.

```
fC <- function(b){ b[1]^2 + b[2]^2 < 1 || (b[1]-2)^2 + (b[2]-0.5)^2 < 0.5 }
bC <- Metropolis( fC, b0=c(0,0), SProp=diag(10,2) )
```

We see that the Metropolis algorithm was able to bridge the gap of zero probability between the two clusters of the distribution. Both clusters are sampled because we used a proposal distribution that very occasionally will propose a jump to the other cluster. So in this case, a proposal distribution that never makes long jumps would have invalidated the algorithm.

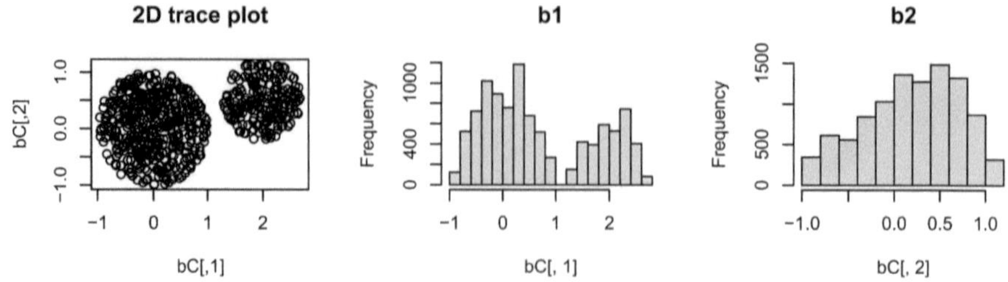

Fig. 6.5 Sampling from an unknown bimodal, bivariate distribution using the Metropolis algorithm (see text)

Exercise

1. **Metropolis for sampling from the posterior.** Take the A-R example at the end of the last chapter, of which the results were shown in Fig. 5.1. Can you replace the A-R with the Metropolis algorithm to sample from the desired posterior distribution? How would the prior and likelihood appear in the Metropolis? (This question makes you think ahead about what will be explained in the next chapter.)

Chapter 7
Sampling from the Posterior Distribution by MCMC

7.1 MCMC and Bayes

In the examples of MCMC in the preceding chapter, no prior or likelihood was specified, nor was there any talk of a posterior distribution. So why is MCMC important for Bayesian analysis?

We have seen that MCMC is a method for sampling from a probability distribution. And as Bayesians, we always want to find a specific probability distribution, namely the posterior distribution. Moreover, when trying to find the posterior distribution, we are often in the situation, required for MCMC, of being able to specify the ratio of probabilities of two parameter vectors, without being able to directly quantify the probability density of any individual parameter vector. That is because Bayes' Theorem tells us that the posterior density for a parameter vector is proportional to the prior density for that vector times its likelihood. So if we can quantify the prior and likelihood ratios, we can use MCMC to sample from the posterior.

MCMC is more efficient for Bayesian calibration than methods such as Accept–Reject that we looked at in Chap. 5 because it does not explore the whole parameter space. Instead it automatically focuses the sampling effort on the region of highest probability; in most circumstances, a chain of 10^4 to 10^5 steps is long enough even for quite parameter-rich models. However, MCMC does require evaluation of the model for each new proposed parameter vector in order to calculate the likelihood. And if the data that are compared with model output in the likelihood function are derived from n different sites (each with different drivers x), then we need to run our model n times at each step. So computational demand still can be considerable, all the more so for PBMs that are run for long time series. One proposed solution has been to include only a fraction of the model's parameters in the calibration (e.g. Minunno et al., 2013b), a practice referred to as *parameter screening*. Parameter screening speeds up calibration by reducing the dimensionality of the parameter

M. van Oijen, *Bayesian Compendium*,
https://doi.org/10.1007/978-3-031-66085-6_7

space that the MCMC needs to traverse. However, it thereby underestimates parameter uncertainty and should if possible be avoided.

Let's now start using MCMC in earnest, for Bayesian calibration of the parameters of a model.

7.1.1 MCMC and Models

Our data are almost always in the form (x, y) where the y are observations and the x are the corresponding conditions such as covariates, coordinates, or times. It is generally easy to analyse such data if we have an analytical model f that can take any value of x as input and spit out a model estimate of the true value directly: $\hat{z} = f(x, \theta)$. But when *time* is an important model input, we often use dynamic models that cannot be solved analytically. We have mentioned the common example of process-based computer models (PBMs). Such models typically require long time series of inputs $\{x(t)\}_{t=1..n}$, which they use to simulate system behaviour iteratively as $z(t+1) = z(t) + f(z(t), x(t), b)$. Such models produce more output than there are data: there are more model input–output combinations $(x(t), z(t))$ than there are observations (x, y). So in Bayesian calibration of PBMs, we can only select the subset of model outputs that corresponds to our observations. The unobserved variables (also called *latent variables*) we only learn about indirectly through the calibrated model.

Because of these differences between analytical models and PBMs, it is hard to write very general code for Bayesian calibration using the Metropolis algorithm. So let's begin modestly, by writing code that works for most analytical models. We shall consider the case of models with multivariate output, such as PBMs, in Chap. 8. Another problem, especially with R, is that it is hard to write code that works both with scalars and vectors and with univariate as well as multivariate model inputs and/or outputs. The code examples in this chapter aim for readability rather than computational efficiency.

7.1.2 The Need for Log-Transformations in MCMC

Probability densities for parameter values that are several standard deviations away from the mean will be very close to zero. And in our MCMCs, we shall often find ourselves walking through parameter space far from the mean of the distribution (whose location we do not know when we start the MCMC). But computers have limited precision to represent very small numbers: typically numbers smaller than about 10^{-300} are rounded to zero. So to avoid numerical problems, we work mostly with log-transformed prior probabilities and log-transformed likelihoods. Let's take the example of the standard normal distribution with mean 0 and standard deviation 1. A standard computer can still represent the probability density $p[\theta] = N[\theta|\mu =$

$0, \sigma^2 = 1]$ when θ is 38 standard deviations away from the mean, but not when it is 39 or more. Densities are then rounded to zero, which makes it impossible to calculate the probability ratios in the MCMC. However, there is no problem when we use $\log(p[\theta])$:

```
dnorm( 38:40, m=0, sd=1 ) ; dnorm( 38:40, m=0, sd=1, log=T)
> [1] 1.097221e-314  0.000000e+00  0.000000e+00
> [1] -722.9189 -761.4189 -800.9189
```

The problem is most severe for parameter-rich models. With 200 independent parameters that are all just three standard deviations from their mean, a computer rounds their joint probability (the product of the 200 densities) to zero, but there are no problems if we calculate the equivalent sum of logarithms instead:

```
z <- rep( 3, 200 ) ; prod( dnorm(z) ) ; sum ( dnorm(z,log=T) )
> [1] 0
> [1] -1083.788
```

So in this chapter, the Metropolis algorithm for MCMC that we introduced in Chap. 6 will be written slightly differently, but mathematically equivalently. We will replace the calculation of the probability *ratio* with the calculation of a log-probability *difference*. Let's begin with an example where we want to use the Metropolis algorithm to sample from the posterior distribution of a simple linear model.

7.2 Bayesian Calibration of a 2-Parameter Model

In this section, we shall be calibrating the straight-line model. So there are two parameters: an intercept and a slope. We have three measurements of variable y which we treat as a function of variable x. For each data point, we specify its independent measurement uncertainty as a Gaussian with mean zero and variance equal to 3 (this should be based on our knowledge of the measurement process). The covariance matrix for the three measurements thus is a diagonal matrix with the value 3 on the diagonal and zeros elsewhere. Here is R-code for these virtual measurements:

```
x  <- c(10,20,30) ; y  <- c(6.09,8.81,10.66)
ny <- length(y)   ; Sy <- diag(3,ny)
```

We also write a short R-function `f()` to represent our model, and we assign a Gaussian prior parameter distribution $N[\mathtt{mb}, \mathtt{Sb}]$. We start with a very large prior uncertainty by setting the 2×2 prior covariance matrix Sb to have values of 10^4 on the diagonal. We also encode the calculation of log(prior) and log(likelihood) as

R-functions, and our final preparation is yet another R-function that calculates the sum of log(prior) and log(likelihood). We call that function `logPost()` to indicate that it calculates the logarithm of the posterior density up to an unknown additive constant. Here is the R-code for these choices:

```
f  <- function( x, b ) { cbind(1,x)%*%b }
mb <- c(0,0) ; nb <- length(mb) ; Sb <- diag(1e4,nb)
logPrior <- function( b, mb.=mb, Sb.=Sb )
  { dmvnorm( b, mb., Sb., log=T ) }
logLik   <- function( b, y.=y, Sy.=Sy, x.=x, f.=f ) {
  dmvnorm( y., mean=f.(x, b), sigma=Sy., log=T ) }
logPost <- function( b, mb.=mb, Sb.=Sb, y.=y, Sy.=Sy, x.=x, f.=f ) {
  logPrior(b, mb., Sb.) + logLik(b, y., Sy., x., f.) }
```

Now we come to the key step: implementing the MCMC. We use the Metropolis algorithm for which we already showed code in the last chapter. But there will be two differences here. One is that our modified function `MetropolisLogPost` works with log-transformations throughout. The second is that the function expects as input the specification of the model `f()` whose parameters `b` we are calibrating. The model is used in the calculation of the likelihood. The fact that we made the model specification one of the inputs for `MetropolisLogPost` makes that function widely applicable: we can use it to calibrate many different models, as we shall see further. But we first stay with the straight-line model.

```
MetropolisLogPost <- function( f.=f, logp=logPost, b0=mb,
                               SProp=Sb/100, ni=1e4, ... ) {
  bChain <- matrix( NA, nrow=ni, ncol=length(b0) ) ; bChain[1,] <- b0
  for( i in 2:ni ) {
     b1  <- bChain[i-1,] + as.numeric( rmvnorm(1,sigma=SProp) )
     if( log(runif(1)) < logp(b1,f.=f.,...) - logp(b0,f.=f.,...) )
          bChain[i,] <- b0 <- b1
     else bChain[i,] <- b0 }
  return(bChain) }
```

After all our preparatory work, it has now become very easy to carry out a Bayesian calibration for the two parameters of our straight-line model. If we accept the default settings for chain length and proposal distribution, we can run the calibration with just this code:

```
bChain1 <- MetropolisLogPost( f )
```

So all we need to do now is examine the chain of parameter values `bChain1` that we have generated. The results are plotted in Fig. 7.1. Unfortunately, these results are not good at all! The trace plots show that the algorithm got stuck frequently for many iterations without accepting new proposed parameter vectors. In other words, the algorithm mixed poorly so that parameter space was not well explored. And

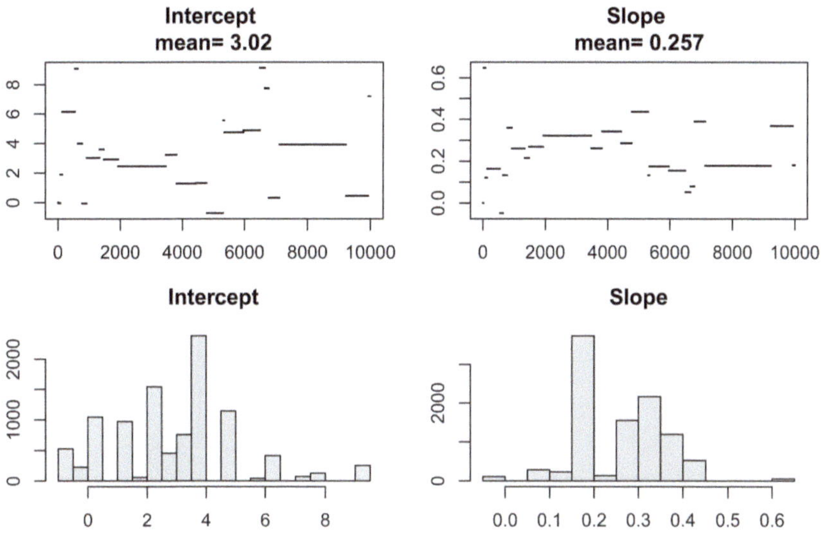

Fig. 7.1 Failed Metropolis sampling of the straight-line model using the default settings for the proposal distribution

the consequence is that our sample from the posterior does not show smooth, bell-shaped histograms. In MCMC-speak: the algorithm has not converged.

So what went wrong? We could say that our chain was not long enough, and the properties of MCMC sampling guarantee that we would get a representative sample from the posterior if we increased the number of iterations manifold. But that is in general not the best response to poorly performing MCMCs. The fact that the vast majority of the 10,000 proposed parameter vectors was rejected means that we should improve our proposal distribution. Remember that we used (our own) default settings for the algorithm, and these included a Gaussian proposal distribution with step length variances equal to 0.01 times the prior variances. But our prior variances were set at the high values of 10^4, so proposal variances were 10^2 and average proposal step lengths were thus about 10. Those step lengths were likely too long to allow the algorithm to properly explore the small part of parameter space that has high posterior probability. So we reduce the proposal step lengths and run the algorithm again:

```
bChain2  <- MetropolisLogPost( f, SProp=diag(c(0.1,0.01)) )
```

Reducing the proposal step lengths makes the algorithm perform much better, as we can see in Fig. 7.2. We now do see that the trace plots are approaching convergence and produce smooth posterior histograms.

Another lesson we can learn from the two MCMCs is the importance of prior information. If we had known beforehand that the intercept would be between 0 and 10 and the slope between 0 and 1, then we could have used smaller proposal steps

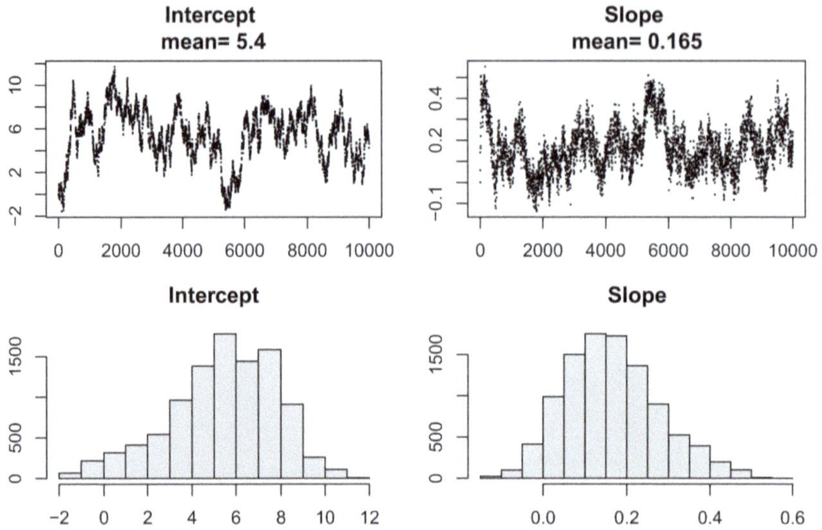

Fig. 7.2 Metropolis sampling of the straight-line model with small proposal variances

from the beginning. But what we have done is a common solution: use a first MCMC to test a default proposal distribution, discard its results and try again with better proposals. There are actually so-called *adaptive MCMC-algorithms* that optimise the proposal distribution on the fly (Andrieu & Thoms, 2008), but their effectiveness depends on the model and data at hand, and we shall not discuss them further in this chapter (some pointers to the literature are given in Chap. 9).

7.3 Bayesian Calibration of a 3-Parameter Model

In this section, we use the same definition of the `logPrior()` and `logLik()` functions, but apply them to a new 3-parameter quadratic model `fq` with its own prior parameter distribution $N[\texttt{mbq}, \texttt{Sbq}]$. We also change the data to make a quadratic model more plausible. The new dataset `yq` will use the same covariate values `x` as before. After making those definitions, we run the Metropolis for this model:

```
x       <- c(10,20,30) ; yq <- y + 0.1*x^2
nyq     <- length(yq)  ; Syq <- diag(30,nyq)
fq      <- function( x, b ) { cbind(1,x,x^2)%*%b }
nbq     <- 3 ; mbq <- rep(0,nbq) ; Sbq <- diag(nbq) * 1e4
bqChain <- MetropolisLogPost( fq, b0=c(0,0,0),
  SProp=diag(c(1,0.1,0.01)), mb.=mbq, Sb.=Sbq, y.=yq )
```

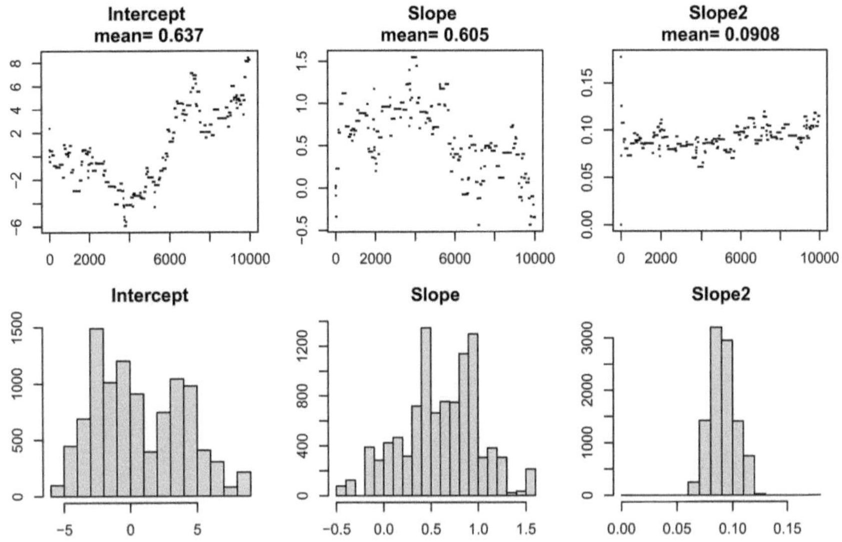

Fig. 7.3 Metropolis sampling of the 3-parameter quadratic model

The results of this Bayesian calibration are shown in Fig. 7.3. There are a few more lessons here. If we study the trace plot for the second slope (the regression coefficient for the quadratic term), then the estimates initially seem to be unstable but gradually they reach values around 0.1 and then the algorithm stays around there. Such an initial phase of instability or drift is standard for MCMC. It arises when we start the chain at an arbitrary point in parameter space (often the mean of the prior) that may be far removed from the area with high posterior probability. So the algorithm needs a number of iterations before it finds the primary area to sample from. That initial drifting phase is called the *burn-in*, and we should remove that part of the chain from our posterior sample because burn-in points are not representative. Looking again at the trace plot, it seems that the burn-in for the second slope parameter is only a few thousand iterations but that is actually wrong! The burn-in is defined for the *joint* probability distribution of all parameters. If one or more parameters still show drift, as the intercept and first slope parameter clearly do, then other parameters that may seem to have stabilised cannot be trusted to stay where they are because parameter estimates are not independent. We must conclude that the MCMC of Fig. 7.3 has not converged yet, so the chain length should be extended.

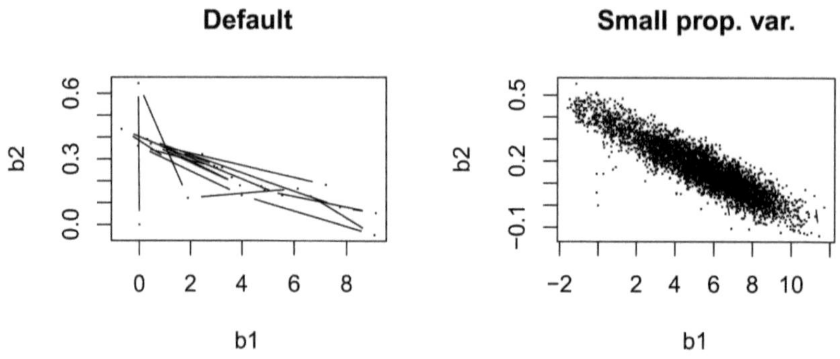

Fig. 7.4 Scatter plots for the two MCMCs of the linear model

7.4 More MCMC Diagnostics

We conclude with mentioning a few more analyses that are good to do after you have carried out an MCMC. You have already seen the value of the trace plots for visually assessing whether the MCMC has converged. (There are also semi-formal methods for quantifying the degree of convergence which we shall discuss in a later chapter.) But remember that the posterior distribution is a joint distribution, so trace plots for individual parameters do not tell the whole story. You can make scatterplots to study posterior correlations between parameters, as shown in Fig. 7.4. Do you recognise the burn-in phase in the panel on the right?

Posterior parameter correlations can of course also be assessed in the form of the posterior covariance or correlation matrix, but those only quantify the strengths of linear relationships, whereas scatter plots can reveal any non-linear interdependencies too.

While you are optimising your proposal distribution, it may be useful to calculate what fraction of the proposals was accepted. But do not aim for excessively high proposal acceptance, because that can be a sign of too small proposal steps and poor mixing. When proposals steps are very small, every proposed parameter vector has nearly the same value for the product of prior and likelihood as the current vector in the chain and is thus accepted with probability close to 1. Practical experience suggests that an acceptance percentage of 20–50% leads to fastest convergence of MCMCs.

7.5 Posterior Predictive Uncertainty

So far, all plots in this chapter have been about the parameters that we are calibrating, but we of course calibrate them to make our models perform better and to quantify predictive uncertainty. As we already noted in Chap. 1, uncertainty

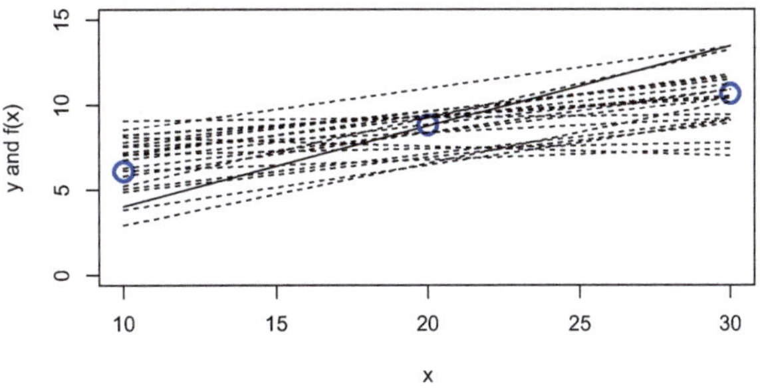

Fig. 7.5 Twenty predictions from the posterior distribution for the linear model

about model inputs (including parameters) propagates to model output, and we should acknowledge this. Sometimes it is easier to recognise calibration problems by studying the posterior predictive distribution than the posterior parameter distribution, e.g. when model output variables are observable and parameters are not. After running your MCMC, you can take an equidistant subsample from the chain, such as the parameter vectors at iterations 200, 400, 600, etc. after burn-in, and run the model for each of these parameter values. Figure 7.5 shows an example produced in this way. It depicts a sample from the predictive distribution generated by our second MCMC (the successful one) for the linear model of this chapter (Fig. 7.2). In this book, we shall not always show the predictive uncertainty if our focus is on parameter estimation, but when reporting research results it should be included (see the reporting checklist in Chap. 10).

In the next chapter, we shall discuss the application of MCMC to complex process-based models. If you then still have questions about MCMC, they may well be answered in Chap. 9.

Exercises

1. **Identifiability of parameters.** Use the same data, likelihood function and prior as before for linear regression, but use a three-parameter model with a clearly redundant formulation: `f <- function(x,b){ cbind(1,x,x)%*%b }`. Note the double appearance of the covariate `x`. Carry out the Bayesian calibration for this model using MCMC and check the sample from the posterior distribution. Are the marginal posterior distributions for the two slope-parameters `b[2]` and `b[3]` the same? Examine the variances for both slope parameters. Why are both variances much larger than the variance of the single slope-parameter in the regular 2-parameter linear model? What is the uncertainty of the *sum* of `b[2]` and `b[3]`? How does that compare to the uncertainty of the slope-parameter in the regular model?
2. **Convergence and posterior.** But what about the MCMC, do you think it will converge more quickly for the 2-parameter model than for the 3-parameter model? To have the MCMCs perform identically, how should the prior parameter

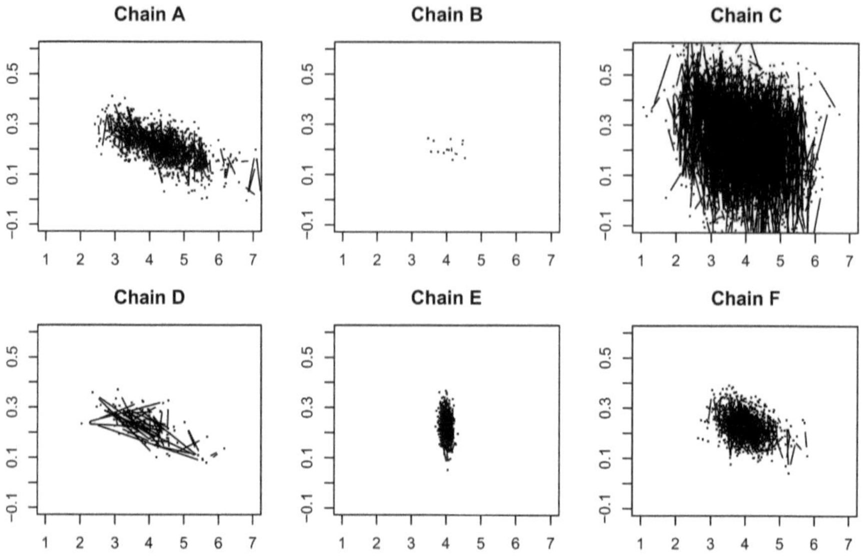

Fig. 7.6 EXERCISE: Try matching panels A-F to MCMCs 1–6 described in the text

distribution and the proposal distribution for the 3-parameter model be set relative to the settings for the 2-parameter model?

3. **Which MCMC produced the following posterior samples?** This exercise is about Fig. 7.6. Each panel in the figure shows the posterior sample from an MCMC. Your task is to provide a correct legend for the figure, i.e. say which panel comes from which MCMC. In all cases, we calibrated the straight-line model as we did earlier in this chapter. But we varied data uncertainty, the prior, the proposal distribution and the number of MCMC iterations. For the purposes of this exercise, the standard settings are `mb` = `c(4, 0.2)`; `Sb` = `diag(1, 2)`; `Sy` = `diag(3, 3)`; `SProp` = `diag(0.1, 2)`; `ni` = `5e3`. The following MCMCs were carried out:

- 1: Standard settings.
- 2: Variance of measurement uncertainty increased by a factor 10.
- 3: Prior variance for the intercept reduced by a factor 100.
- 4: Prior variances for both parameters decreased by a factor 5.
- 5: Proposal variance increased by a factor 50.
- 6: Number of MCMC iterations reduced by a factor 100.

Now try to match figure panels A-F to these MCMCs 1–6. We give one answer already: the first panel shows the results from using the standard settings. So panel A = MCMC 1.

Chapter 8
MCMC and Multivariate Models

In this chapter, we focus on models with multivariate output. That includes most process-based models (PBMs). Models with multivariate output are not fundamentally different from the simpler models we studied in the previous chapters; we can still write them as functions f of their input consisting of covariates x and parameters θ. But the output $z = f(x, \theta)$ from the models will be multivariate, e.g. time series of different properties of an ecosystem. That does not affect the principles of Bayesian calibration in any way but may complicate its execution. In this chapter, we illustrate these issues with a quite simple PBM that as output produces two time series: the growth over time of the biomass and leaf area of vegetation.

8.1 Process-Based Models (PBMs)

Process-based models (PBMs) are multivariate functions. They are dynamic models that can usually not be analytically solved, so to calculate their output we would have to use numerical integration, iterating over time steps. The input required by PBMs typically consists of time series of driver variables (the covariates x), and parameters θ which include the initial values of the *state variables*. The iterative process of solving a PBM starts from these initial values, and then alternates between calculating rates of change and updating the corresponding state variables.

8.1.1 A Simple PBM: The Expolinear Model

The model that we use in this chapter is the 5-parameter expolinear model, `EXPOL5`, which was developed and studied by Goudriaan and Monteith (1990). It can be

M. van Oijen, *Bayesian Compendium*,
https://doi.org/10.1007/978-3-031-66085-6_8

written as a single differential equation:

$$\frac{dW}{dt} = LUE \cdot I_0 \, (1 - e^{-K \cdot LAR \cdot W}); \quad W(0) = W_0, \tag{8.1}$$

where W is biomass (g m^{-2}), LUE is the Light-Use Efficiency (g MJ^{-1}), I_0 is light intensity (MJ m^{-2} d^{-1}), K is the light-extinction coefficient (m^2 m^{-2}), LAR is the Leaf Area Ratio (m^2 g^{-1}), and W_0 is the biomass at $t = 0$. Despite its brevity, this equation represents considerable knowledge about vegetation growth. The product of LAR and W is the so-called leaf-area index (LAI) of the vegetation, which is the primary variable that determines resource use. The exponential term represents *Beer's Law* which quantifies the fraction of light that is transmitted through a medium, so the full bracketed term represents the fractional light interception by the canopy. Multiplied with the incident light intensity, I_0, we arrive at the amount of light intercepted by the vegetation per unit of time and ground area. Finally, the multiplication with the LUE represents the process of photosynthesis corrected for respiration, giving us the vegetation growth rate $\frac{dW}{dt}$. The equation has shown to be suitable to describe the growth from seedling to maturity of many different plant species growing in monoculture.

Remarkably, this sophisticated one-equation process-based model allows for an analytical solution if the light intensity remains constant over the period of interest:

EXPOLINEAR EQUATION:

$$W(t) = \frac{\ln\left(1 + \exp(I_0 \cdot K \cdot LAR \cdot LUE \cdot t)\,(\exp(K \cdot LAR \cdot W_0) - 1)\right)}{K \cdot LAR} \tag{8.2}$$

You can verify that $W(0) = W_0$. We now want to write an R-function for the expolinear equation to facilitate MCMC sampling. We shall be having data for both biomass W and leaf-area index LAI, so we make those two variates the outputs from our function EXPOL5:

```
EXPOL5 <- function( t=0, b=c(10,1,0.007,2,1) ) {
  I0  <- b[1] ; K <- b[2] ; LAR <- b[3] ; LUE <- b[4] ; W0 <- b[5]
  W   <- log( 1 + exp(I0*K*LAR*LUE*t) * (exp(K*LAR*W0)-1) ) / (K*LAR)
  LAI <- W * LAR
  return( list( "W"=W, "LAI"=LAI ) ) }
```

The "5" at the end of the R-function name 'EXPOL5' is to emphasise that this is a 5-parameter model, with the parameter vector representing (I_0, K, LAR, LUE, W_0). Our prior for the parameters of EXPOL5 shall be a multivariate Gaussian with mean parameter vector mb and covariance matrix Sb:

```
bmin5  <- c(9.9,0.5,0.005,1,0.5)            ; bmax5 <- c(10.1,1.5,0.009,3,1.5)
range5 <- bmax5 - bmin5                      ; var5  <- (range5^2) / 12
mb     <- rowMeans( cbind(bmin5,bmax5) ) ; Sb    <- diag( var5 )
```

8.2 Bayesian Calibration of the Expolinear Model

We want to demonstrate a Bayesian calibration of the parameters of the expolinear model, and here are the (virtual) data that we shall be using:

```
t_W      <- c(20,60,100) ; y_W <- c(10,500,1200) ; sd_W <- c(5,15,120)
t_L      <- c(20,80,100) ; y_L <- c(0.2,5,4)     ; sd_L <- c(0.1,1,1)
data_W   <- cbind( t=t_W, y=y_W, sd=sd_W )
data_LAI <- cbind( t=t_L, y=y_L, sd=sd_L )
data     <- list( "W"=data_W, "LAI"=data_LAI )
```

We shall be using the same function for the logarithm of the prior as in Chap. 7, but we need to define a new R-function for the log-likelihood that can handle the kind of output list for multiple variates that EXPOL5 produces:

```
logLikList <- function( b, dataList, f.=f ) {
  sum( sapply( names(dataList), function(v){ sum(
    dnorm( f.( dataList[[v]][,1], b ) [[v]],
           mean=dataList[[v]][,2], sd=dataList[[v]][,3], log=T ) ) } ) ) }
```

As you can see from the code, this likelihood function treats all measurements of the two output variables as independent and as having known datapoint-specific measurement uncertainty (the values of the elements of sd_W and sd_L in our virtual dataset). When you calibrate your own model, you must decide whether that approach is suitable for your case too, or whether measurement uncertainties need to be calibrated rather than prescribed. It will depend on your knowledge of the measurement process.

Because we are using a different likelihood function (logLikList()) than in the previous chapter (where we used logLik()), we also need a slightly adapted version of the function for the sum of log(prior) and log(likelihood):

```
logPostList <- function( b, mb.=mb, Sb.=Sb, dataList, f.=f ) {
  logPrior(b, mb., Sb.) + logLikList(b, dataList, f.) }
```

Now we have everything in place for Bayesian calibration of our process-based model, EXPOL5. We run the exact same R-function MetropolisLogPost that we defined in Chap. 7:

```
bChain <- MetropolisLogPost( f=EXPOL5, logp=logPostList, dataList=data )
```

The five parameter trace plots and parameter histograms are shown in Fig. 8.1.

Note that we included calibration for an environmental driver, namely the radiation parameter I_0. That is perfectly fine: environmental drivers are just one class of parameters or parameter vectors and can be treated in the same way as the other model parameters. This means that from a Bayesian perspective there is never

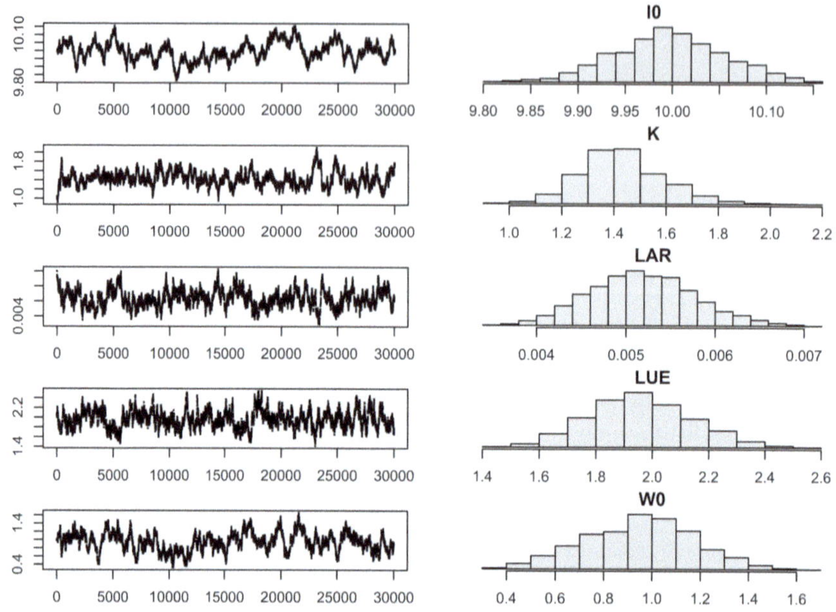

Fig. 8.1 Bayesian calibration of the 5-parameter expolinear model (EXPOL5) by means of Metropolis sampling

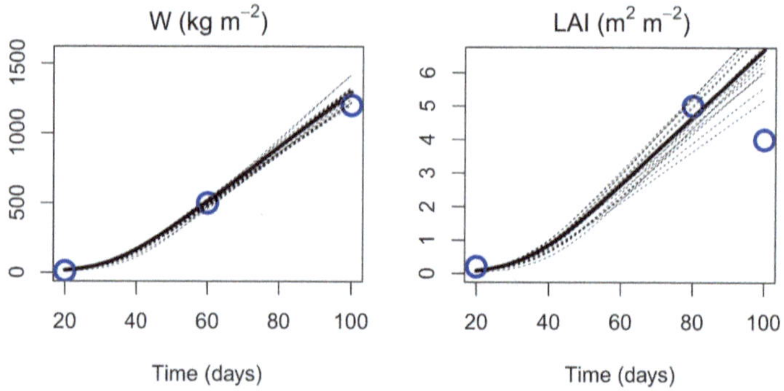

Fig. 8.2 Time series of biomass (W) and leaf-area index (LAI), generated by the expolinear equation using the posterior mean parameter vector (thick line) and 20 parameter vectors subsampled from the MCMC (thin dashed lines)

a need for ad-hoc driver gap-filling algorithms. Gap filling is a parameter-estimation problem like any other and can be addressed with the same Bayesian calibration approach. Also note that we calibrated the initial value W_0 of state variable W which is a parameter too.

Finally, we make plots of EXPOL5 outputs vs. data to verify that the posterior parameter distribution makes sense (Fig. 8.2).

8.3 More Complex Models

This chapter showed that the Metropolis algorithm and MCMC algorithms in general really are Swiss army knives that can be used for Bayesian calibration of any dataset and model, irrespective of the number of variates involved. Bayesian calibration of complex models is gradually becoming more common (see e.g. the review by Van Oijen, 2017). If you want to explore such models: some can be freely downloaded from web repositories. The following three examples of downloadable process-based vegetation models come with code for Bayesian calibration, in each case by means of the Metropolis algorithm:

- `BASGRA_N`, a model for grasslands (Van Oijen et al., 2018),
- `BASFOR`, a model for forests (Van Oijen & Cameron, 2020),
- `CAF2021`, a model for coffee agroforestry systems (Van Oijen, 2022).

The following chapter will discuss a variety of issues surrounding the use of MCMC that we have partly skipped over so far.

Exercise

1. **Parameter correlations.** Figure 8.1 depicts the marginal posterior probability distributions for the five parameters of `EXPOL5`. So it gives us a good idea of posterior uncertainty for each individual parameter. But it is an incomplete picture of the joint parameter distribution. Run the R-code of this chapter (which should take just a few seconds) and then quantify all pairwise parameter correlations with one line of R-code that calls the function `cor()`. Are there parameters that are strongly correlated, say with an absolute correlation coefficient greater than 0.7?

Chapter 9
Bayesian Calibration and MCMC: Frequently Asked Questions

This chapter answers a number of common questions about Bayesian calibration in general and MCMC in particular. Many topics are addressed elsewhere in this book at greater length (and pointers to the chapters are then given), but some are only addressed briefly here.

9.1 The MCMC Algorithm

1. Can the use of more than one MCMC chain help assess convergence? Yes. If you run more than one chain, each starting from a different point in parameter space, it is much easier to see when the chains have converged. Also, the individual chains may then be shorter, and if you can run the chains in parallel, computational demand will not increase much (Margossian & Gelman, 2024). Convergence is apparent when the different chains have moved to the same part of parameter space—and we can establish this formally by testing whether the interchain variance has come down to the level of intrachain variance using, for example, the Gelman–Rubin statistic (Gelman et al., 2013) or modern variants thereof (Vehtari et al., 2021). However, full-proof convergence diagnostics do not exist, and visual inspection of trace plots remains important.

2. My model is a process-based model: does that affect how I should carry out Bayesian calibration? In principle, not. Bayesian thinking is completely generic and applies to all models and datasets. However, MCMC requires many runs of your model, so if your model is computationally demanding, consider replacing it with a fast emulator (see Chap. 15).

M. van Oijen, *Bayesian Compendium*,
https://doi.org/10.1007/978-3-031-66085-6_9

3. My model is perhaps not very good: does that affect how I should run the MCMC? No. The calibration by MCMC takes your model as given. The posterior distribution for the parameters is always conditional on both the data and the model that are being used. We could express this by writing $p[\theta|y, f]$ for our posterior distribution, rather than just $p[\theta|y]$. However, we normally write the shorter version and treat the conditionality on the model as understood. Note that the use of a poor model together with informative data may lead to a posterior distribution for the parameters that will seem unrealistic if considered in isolation—even though conditional on the model that posterior is the correct one. The Bayesian calibration will thus help us identify models of poor quality—and in such cases, it is advisable to try different models and carry out a Bayesian Model Comparison (BMC; see Chap. 11). There are methods for Bayesian calibration which do not assume that the original model is correct. Instead they add a stochastic term to the model, called the 'discrepancy', to represent the mismatch between model behaviour and reality. Often the discrepancy is modelled as a multivariate Gaussian that quantifies the mismatch for each data point. In short, these methods replace the assumption of the original deterministic model being correct with that of the extended (deterministic + stochastic) model being correct. A problem with that approach is that the stochastic discrepancy term is only defined well near data points and does not help us much with predictions. We discuss the discrepancy at greater length in Chap. 12.

4. Are non-MCMC methods good alternatives? Non-MCMC methods such as Accept–Reject (A-R; Chap. 5) should not be used with parameter-rich models such as PBMs, unless a screening step is added in which the number of to-be-calibrated parameters is reduced to a small number. A-R has problems with parameter-rich models because it begins by sampling from the prior across the whole of parameter space. This cannot be done exhaustively if the space is high-dimensional. In contrast, MCMC samples directly from the posterior distribution and is efficient in that it spends most time in the area of high probability. However, A-R and related non-MCMC methods such as Importance Sampling have advantages too: they are simpler and easier to apply because there is no proposal distribution to optimise.

5. Which MCMC algorithm should I use? Metropolis (Chap. 6) and Gibbs Sampling (Chap. 14) are completely general: they will give the right answer for any combination of model and data—provided the sample from the posterior is large enough. However, there is a multitude of different MCMC algorithms, and new ones keep being developed. Many are adaptive MCMC algorithms in which the proposal distribution changes during the random walk (Andrieu & Thoms, 2008). The adaptation method must be carefully designed to guarantee that the algorithm converges to the correct posterior distribution (Robert & Casella, 2010). Having said that, the future is likely to see increasing use of adaptive algorithms. Methods that have shown to be useful include DRAM (Haario et al., 2006) and DEMCzs (ter Braak & Vrugt, 2008). Of the algorithms for which examples are included in this book, the Metropolis is the one that works best for parameter-rich models such as process-based models (PBM), say more than 10 parameters to be calibrated. Gibbs sampling (Chap. 14) generally cannot be implemented for PBMs because the conditional probability distributions for parameters cannot be defined. Similarly unsuitable for PBMs are gradient-based methods, such as Hamiltonian MC (HMC), because these require the prior and likelihood function to be differentiable (Betancourt, 2018; Štrumbelj et al., 2024). However, HMC and especially adaptive versions of it such as NUTS (built into the Bayesian calibration software Stan) are increasingly popular because they are very efficient tools for simpler models.

9.2 Data and Likelihood Function

See also Chap. 4.

1. Does it matter whether we use all data at the same time or in steps? No. Assume you have two datasets y_1 and y_2. You could use all the data at once to calculate the posterior $p[\theta|y_1, y_2]$. Alternatively, you could first calculate $p[\theta|y_1]$ and then use that as the prior for a Bayesian calibration on y_2. In both cases, you arrive at the same final posterior distribution $p[\theta|y_1, y_2]$. This is formally shown in Appendix C (*Probability Theory for Modellers*) under the heading *Sequential Bayesian Updating*, and it shows one of the major strengths of Bayesian calibration: it is firmly based on probability theory and always gives the same answer if the same information (in this case y_1 plus y_2) is used.

2. How do I handle systematic measurement error? Systematic measurement error, where all measurements are off by a constant but unknown amount or proportion, can be handled like any other unknown constant: as a parameter that we calibrate. This only tends to be necessary for some types of measurements, e.g. poorly replicated ones. An example may be the measurement of soil emissions of N_2O, where we are always in danger of unrepresentative placement of soil chambers. For such emission data, we could use a multiplicative systematic error parameter (for an example of this approach, see Van Oijen et al., 2011).

3. I have a lot of data: do I need to use them all? In principle, yes, as long as the uncertainty about random as well as systematic measurement errors is properly quantified in the likelihood. If the model is very good, then every bit of empirical information will tend to give us a posterior distribution for the parameters which will improve model performance for any application. However, with our imperfect models, it may be prudent to aim for balance between different kinds of measurement information. We can do this by not collecting many more data for one variable, one site or one time period, than for others (Cameron et al., 2022).

4. What do I do if measurement errors are not independent? This is important for calculating the likelihood. We often assume that measurement errors are independent and calculate the overall likelihood of a dataset $p[y|\theta]$ as the product of likelihoods for the individual data points $p[y_i|\theta]$. This is usually done as a straightforward multiplication of Gaussian probability densities, with the standard deviations σ_i representing each data point's uncertainty. If the measurement errors are correlated we can take that into account by calculating the likelihood differently, e.g. as a multivariate normal distribution where the off-diagonal elements of the covariance matrix are not all zero (Chap. 4). For technical details on this and other methods for modifying the likelihood function, see, for example, Gregory (2005) or Sivia and Skilling (2006).

5. What do I do if my measurements have uncertainties in the x- as well the y-direction? This can happen if it is unclear at what exact time the measurements were taken. This is easily incorporated in the Bayesian calibration, by adjusting the calculation of the likelihood. Say there is 50% chance that a certain data point $y(t)$ was indeed measured at the reported day and 25% chance each for the days before and after. Then the likelihood for that point is calculated as the weighted average of the likelihoods calculated for each of the three days, with weights 0.25, 0.5 and 0.25.

9.3 Parameters and Prior

See also Chap. 3.

1. Can Bayesian calibration handle correlations between parameters? Yes. Correlations appear whenever we do not know the exact values of two parameters but do know something about their combined values. For example, we may know that if parameter θ_1 has a high value, parameter θ_2 is likely to have a low value. In fact, each time we speak of "the" parameter distribution, we speak of one single probability distribution for all possible combinations of parameter values, and that distribution typically includes correlations. In other words, the parameters' distribution is a *joint* probability distribution, not a collection of independent distributions for individual parameters. Furthermore, even if the prior distribution has no correlations between parameters, the posterior distribution almost always will.

2. Should I include all my model's parameters in the Bayesian calibration or just a subset? Preferably all of them. Using a subset of the parameters will inevitably lead to underestimation of the contribution of parameter uncertainty to predictive uncertainty. MCMC can handle high-dimensional parameter vectors, but the A-R method will struggle. However, if it is possible to identify a subset of parameters that together account for most of the model's behaviour, then there is little harm in using the subset only. That practice is called 'parameter screening'. Some models are so parameter-rich that screening is inevitable (e.g. global climate models, and neural networks used for natural language processing).

3. What distribution should I use for the prior distribution? The prior distribution should represent what we know about our model's parameters. That means representing what parameter values are considered impossible, which ones are possible but not likely, which ones are quite plausible, etc. If we have no idea about parameter correlations, our prior can simply be the product of different univariate distributions that we assign to individual parameters. Using the uniform distribution is generally not a good idea: we typically do not consider all values between some lower and upper limits equally plausible. A triangular distribution is better but still has strange discontinuities. Good choices are the (truncated) normal distribution, the lognormal distribution and the beta distribution. However, there is no golden rule: it all depends on what we can say a priori about our parameter values. Note that we may want to represent, in the prior, complicated constraints on the parameter values. For example, we may have two parameters θ_{min} and θ_{max} that are both highly uncertain but we know that θ_{min} has to be smaller than θ_{max}. The easiest way to include such information in the prior distribution for the parameters is to reparameterise, e.g. by replacing θ_{min} with a new parameter k_{min}, defined as $\theta_{min}/\theta_{max}$. Reparameterisation may lead to parameter vectors that are easier to calibrate (Gilks et al., 1995).

9.4 Code Efficiency and Computational Issues

1. How do I make the proposal algorithm of my MCMC method more efficient? If proposal acceptance rate is low, the average step length of the MCMC may be too long, and vice versa. Acceptance rates around 20–50% often give best convergence. It may also be that the step lengths are OK but that we are making too many steps in the wrong direction. This could happen, for example, if we use a radially symmetric proposal distribution while the posterior that we are aiming for has strong correlations between parameters, i.e. is ridge-shaped. Besides using a different MCMC algorithm (e.g. an adaptive one, see the first section of this chapter), you can begin by performing some trial-and-error calibration optimisation to find an efficient proposal distribution. Gelman et al. (2013) give some suggestions for optimising the proposal in algorithms like Metropolis that use a multivariate normal distribution to generate the random steps. They advise to make the proposal distribution proportional to the posterior distribution that we are going to generate. However, we do not know this distribution beforehand—that is why we do the MCMC in the first place—so we initially have to work with the prior distribution instead. Some degree of trial-and-error remains necessary. The most general recommendation that can be made is to try out various proposal distributions in short chains and then select the best one for the definitive Bayesian calibration.

2. What do I do if the MCMC crashes? Obviously, this is not supposed to happen! Fortunately, it is also very unlikely given the simplicity and robustness of most MCMC algorithms. However, there is one part in the algorithms that occasionally gives problems: calculation of the prior probability for a new candidate parameter vector. If the prior distribution is very simple, e.g. multivariate uniform or triangular, calculation of the prior probability density for any new parameter vector poses no problems. However, if we use a distribution like the multivariate normal for the prior, then calculation of parameter vector probability includes a matrix inversion step, where the covariance matrix needs to be inverted. This may run into limitations with the numerical precision of computers. Matrix inversion is a widely studied issue in applied mathematics, and there are robust but complicated algorithms available. However, we can generally avoid all problems if we don't have too many orders of magnitude difference between the smallest and largest parameters. A large covariance matrix where some entries are in the order of 10^{-6} and others 10^8 is difficult to invert. We can deal with this by rescaling the model's parameters to be all close to unity. Of course, that requires undoing the rescaling each time we need to run our model.

3. What do I do if the MCMC does not converge? The first thing to do is make the chain longer. If that takes too much computing time, consider changing the proposal algorithm, e.g. by making the proposal distribution narrower (i.e. shortening the average proposal step). You can of course also try a different MCMC algorithm altogether.

4. My PBM is too slow for Bayesian calibration because it needs a long spin-up, what do I do? Your model does not *need* a spin-up. Spinning-up, which is pre-running the PBM to equilibrium, is used to initialise the state variables of a dynamic model. This is done to prevent unrealistic model dynamics when the true runs start. After spinning-up, the model does not need to find its way back to stable behaviour. However, this generally amounts to replacing initially unrealistic system dynamics with unrealistic initial values of state variables, and this cannot be recommended. Yeluripati et al. (2009) showed a superior initialisation approach where soil measurements were not discarded but used together with flux-data in a Bayesian calibration of the parameters and initial values of a model for vegetation gas exchange. This way, spin-up was avoided.

9.5 Results from the Bayesian Calibration

1. What is burn-in and do I need to consider it? Burn-in is the first phase of an MCMC, when the chain is still "searching" for the area of high posterior probability, i.e. before convergence. The parameter vectors sampled during the burn-in phase should be removed from the posterior sample because they are not representative.

2. If my chain length is, say, 10,000 and acceptance rate is 25 %, does that mean my sample from the posterior only consists of 2500 parameter vectors? No, the sample consists of all 10,000 parameter vectors, but many vectors will appear two or more times in the sample. In other words, if during the MCMC, a candidate parameter vector is rejected, the current vector will receive extra weight. So if we want to calculate, for example, the mean posterior parameter vector, we need to sum all 10,000 vectors (including the repeats) and divide by 10,000.

3. What do I do if the sample from the posterior parameter distribution looks wrong, or if the posterior predictions of model output variables look wrong? Not necessarily anything. All the Bayesian calibration does is combine the information that has been supplied—via the prior, the model and the data-likelihood—and show what that means for what we can say about the parameters. The results may be surprising but nevertheless correct. However, if something really seems amiss, e.g. posterior predictions for some variables seem impossible, there may be an error somewhere. First check the MCMC code. Then check if the prior was incorrectly set, e.g. by disallowing perfectly reasonable parameter values. Then check whether the likelihood was correctly defined: perhaps there was less information in the data than you assumed, e.g. if you neglected representation error (where the data were accurate but not representative of what the model actually tries to simulate) or if you overlooked systematic errors in the data. It may also be that you have lots of unused information: if you have good reason to disbelieve certain posterior predictions, then that suggests you have information that could be included in the likelihood function. What's left after all those checks is the most common cause of seemingly poor Bayesian calibration results: the model itself may not be very good and the calibration just makes that very apparent. In that case there are four options left: (1) Accept your results but consider them to be conditional on a model that you think is flawed; (2) improve your model; (3) re-interpret your model as not being intended to simulate all the characteristics of the data, e.g. by treating your model outputs as being correct for a specific spatiotemporal scale, which may have lower resolution than that of the data; (4) redo the MCMC with a discrepancy term included (Chap. 12).

4. How do I summarise my results? Try not to summarise too much: your results are represented by the whole posterior distribution and the associated predictive distribution (see Chap. 10 for more on this). You can use barplots or other graphical means to show the distributions. But if you have to be succinct, some key characteristics of the posterior are its mode, mean and covariance matrix. The mode is the vector in the posterior sample with the highest probability, i.e. the highest product of prior and likelihood. It is also called the *Maximum A Posteriori parameter vector* (MAP). The MAP will be equal to the mean if the posterior is symmetric. When you have many data, the posterior is often well approximated by a multivariate Gaussian distribution defined by the sample mean and the covariance matrix. But if the posterior has a more complicated shape, advanced *density estimation* techniques may be required. Approximating the posterior sample by a Gaussian or other parameterised distribution (i.e. density estimation) is essential in case new data become available with which we want to do a further Bayesian calibration. The posterior from the first Bayesian calibration then becomes the prior for the second Bayesian calibration. We need a parameterised distribution for this new prior to allow us to calculate the prior probability density for any candidate vector in parameter space.

Chapter 10
After the Calibration: Interpretation, Reporting, Visualisation

This chapter discusses what needs to be done after your Bayesian calibration: how to interpret your results, what to report and how to report it with emphasis on visualisation.

10.1 Interpreting the Posterior and Model Diagnostics

A major benefit of Bayesian calibration of the model's parameters is that it helps you find structural errors in the model. After calibration, any remaining differences between observations and model outputs will be mainly due to data error and model structural error. It is impossible to fully disentangle these two errors, but some heuristic advice can be given (Van Oijen et al., 2011). We suggest the following five steps, some of which are especially useful for complex process-based models (PBMs):

1. The first step after Bayesian calibration is to inspect the marginal posterior probability distributions, i.e. the probability distributions for each individual parameter in the model. When any of these probability distributions is highly skewed, creeping up to its lower or upper prior bound, then the data contradicted our prior expectations, and we need to examine whether there were any errors in prior, data or the model itself. Often it will be the latter—it may be an indication that the processes affected by that parameter are incorrectly implemented. For example, if the posterior litter decomposition rate constant in a soil model is higher than expected, then it may be the case that other decomposition pathways were overlooked.
2. The second step is to inspect correlations between parameters. Strong posterior correlations between parameters of any model indicate that some function of the parameters can, a posteriori, be estimated more accurately than the individual parameters. A trivial example is where the model is simply the sum of two

M. van Oijen, *Bayesian Compendium*,
https://doi.org/10.1007/978-3-031-66085-6_10

parameters ($y = \theta_1 + \theta_2$) and where we have one observation of the sum. Then the calibration will induce a strong negative correlation between the parameters, so each individually will be more uncertain than their sum. Other functions of the parameters, such as their difference, may remain highly uncertain (see also Jaynes, 2003, p. 267). The presence of strong correlations between parameters may indicate that the model is more complex than it needs to be.

3. The third step is decomposition of the mean squared deviation (MSD) of time series (Kobayashi & Salam, 2000). This is of relevance for dynamic models such as PBMs, which, for example, simulate the dynamics of demographics or biogeochemistry. After the calibration, we can compare observed and simulated time series. The MSD decomposition proposed by Kobayashi and Salam (2000) splits the MSD into three additive terms that quantify differences between the time series in their mean, variance and phase:

$$MSD = \overline{(z-y)^2} = (\bar{z}-\bar{y})^2 + (\sigma_z - \sigma_y)^2 + 2\sigma_z\sigma_y(1-\rho),$$

   where y refers to observations and z to model output, σ is standard deviations, and ρ is the correlation coefficient. If model and data mean differ (the first term), the model may be missing a process. If the variances differ (second term), the model may be missing a feedback mechanism. If the phases differ (third term), the model may be missing a linked state variable, e.g. in a decomposition chain. This MSD decomposition was applied after Bayesian calibration to the mismatch between observed and simulated gas flux time series for four forest models calibrated for a Norway spruce site in Germany (Van Oijen et al., 2011), to two models for growth and respiration of *E. globulus* in Portugal (Minunno et al., 2013a) and to grassland simulations in Germany (Hjelkrem et al., 2017).
4. The fourth step is to check any assumptions of parameter universality. Did we carry out a calibration using data from multiple geographical locations assuming that parameter values were generic, i.e. the same everywhere? Then we could redo the calibration with genotype- or site-specific parameter estimation. However, this approach pits two extreme ideas against each other: generic vs. site-specific parameters. A better approach would be to allow parameters to vary between sites or genotypes but not completely independently—and let the calibration reveal the appropriate degree of parameter independence. Such a more flexible, hierarchical approach is becoming common in environmental and ecological modelling (Cressie et al., 2009) and will be discussed in Chap. 17.
5. The fifth step is assessing the sensitivity of model output to the individual parameters. This is usually called *sensitivity analysis* (SA) when each parameter is changed by the same percentage, and *uncertainty analysis* (UA) when each parameter is changed according to its uncertainty (e.g. by changing parameters to specific quantiles of their marginal posterior distribution). The parameter changes in SA are arbitrary, so UA is the more useful technique. UA may help identify parameters that we should measure more precisely. But it is not useful for unobservable (hidden, latent) parameters that we can only quantify

by model calibration. UA is most appropriate for models where the parameters have a physical meaning, such as PBMs. Van Oijen et al. (2016) used UA of a grassland PBM to identify the physiological properties of grasses that a plant breeder should focus on. UA may also help with model diagnostics: if our system understanding tells us that certain parameters are important, but our analysis tells us that model output hardly depends on them, then we may need to revise the part of the model in which the parameter occurs. For a thorough overview of techniques used in SA and UA, see Saltelli (2008).

10.2 Reporting

When writing a publication involving Bayesian calibration, we should report on every step. To remind you of these, here is a short checklist that you could use:

1. The data
   - What data did you use? What were the key observational uncertainties?
2. The model
   - What model did you use? If too large for detailed description, where can we find more information about it? How many state variables and parameters does it have?
3. The prior
   - Did you calibrate all parameters or just a subset? What joint probability distribution did you assign? What information did you use to quantify it: literature, personal judgment, expert elicitation? Did you use MaxEnt?
4. The likelihood
   - What likelihood function did you assign? How did you account for stochastic, systematic and representativeness errors? Did you treat the model as given or did you include a discrepancy term (Chap. 12)?
5. The MCMC
   - Which algorithm did you use? How many chains and iterations? How did you assess convergence?
6. The posterior distribution for the parameters
   - What were posterior modes, means, variances and major correlations? How different was the posterior from the prior?

7. The posterior predictive distribution
   - How well did the model, when using the posterior probability distribution for the parameters, (a) reproduce data used in the calibration, (b) predict data not used in the calibration?

Our theme in this book has been that uncertainties are represented by probability distributions, and the checklist above asks to report on them. This raises the issue: how to communicate probability distributions to others in, for example, a scientific publication or a report for management? This is a complicated issue, exacerbated by the fact that the posterior predictive uncertainty for different output variables is typically represented by a high-dimensional joint probability distribution with many non-zero correlations.

Formally, only the full joint probability distribution captures all information about predictive uncertainty. However, summarisation is inevitably required, and the way to report uncertainty should depend on both the shape of the distribution and the needs of the persons who are expected to read your report (Spiegelhalter et al., 2011). If the distribution is symmetrical and bell-shaped, then providing the mean value and the standard deviation may be sufficient. For more complicated distributions, reporting quantiles may be preferable and strong correlations should be flagged (see e.g. Van Oijen et al., 2005).

10.3 Visualising Uncertainty

We note that the field of uncertainty communication is still in flux, and several researchers recommend the use of different graphical representations of uncertainty over verbal or numerical reporting (Gabry et al., 2019; Milne et al., 2015; Spiegelhalter et al., 2011). Examples of reporting posterior uncertainties following Bayesian calibration of vegetation models are provided by, amongst others, Van Oijen et al. (2013b, 2011, 2005) and a map showing how predictive uncertainty of forest growth varies spatially across the United Kingdom is provided by Van Oijen and Thomson (2010).

There is no single right way to plot probability distributions. You need to consider your message and your audience (Hullman et al., 2019). Take the four examples shown in Fig. 10.1. From left to right, they show more information about the distribution, and this book never uses the two on the left. In interviews conducted by Milne et al. (2015), research scientists tended to express a preference for boxplots over density strips, but people from other professions preferred the density strips, especially when combined with verbal explanations.

In its simplest form, the Bayesian method formalises how information flows from data to model parameters and ultimately model output. In later chapters, we shall see that the method can be expanded to a much more comprehensive framework involving multiple models, model emulators, uncertain drivers, uncertain parameters and hyperparameters combined in a Bayesian hierarchical model. Whilst

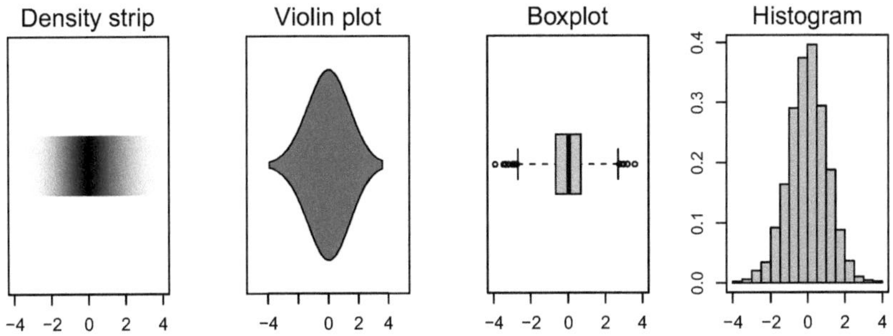

Fig. 10.1 Four ways of plotting the same distribution. Which one do you prefer?

powerful, the complexity of the framework may obscure a reader's view of the information flows between the different components. In this book, we generally opt for verbal descriptions, mathematical equations and computer code to explain the methodology, but a visual alternative exists in the form of graphical models (GM, e.g. Smith et al., 2012). GMs typically depict all uncertain quantities in separate circles, linked by arrows representing conditional probabilities. For example, an arrow from a circle labelled θ to one labelled y would represent the likelihood $p[y|\theta]$ (see Chap. 16). The use of graphical models is a means of visualising the Bayesian approach to uncertainties, but equations remain essential to communicate the exact details of the approach. In any case, many different types of graph can be used. The online R Graph Gallery (https://www.r-graph-gallery.com/) provides hundreds of examples.

As a final note, this chapter is titled 'After the calibration ...' but visualisation can of course help you define likelihoods and find inconsistencies in your prior as well. Gabry et al. (2019) advocate standardising visualisation across the whole Bayesian workflow.

Chapter 11
Model Ensembles: BMC and BMA

In this chapter, we discuss how multiple 'competing' models can be used simultaneously. There are advantages to having more than one model, as was already recognised by Chamberlin in the nineteenth century (Chamberlin, 1890). His still highly readable and important essay on 'The method of multiple working hypotheses' warned scientists against 'parental affection for a favourite theory'. He worried that inevitable bias in favour of one's own ideas would lead to 'unconscious pressing of the theory to make it fit the facts, and a pressing of the facts to make them fit the theory'. And that is—or should be—a legitimate concern for modellers nowadays as well. Moreover, different models can have complementary strengths, and we often have no clear idea which of the available models is the best for a given research question. So how can Bayesian thinking help with these issues? Well, as you will expect, the proper Bayesian approach is to quantify our uncertainty about model structure. One way to do that is by following Chamberlin's advice to use multiple models.

11.1 Ensembles, Integrated Likelihoods and Bayes Factors

Uncertainty with respect to model structure itself is more difficult to represent formally than parameter uncertainty. For parameters, as we have seen, we can simply define a probability distribution $p[\theta]$, but how would we specify a probability distribution $p[f]$? A practical solution to this issue is to collect multiple models and assume that the set forms a representative sample from 'model space'. Such *ensemble modelling* is becoming increasingly common in the environmental sciences (Chandler, 2013). Examples using vegetation models are provided by Van Oijen et al. (2013b) and Rollinson et al. (2017). In ensemble modelling, we define $p[f]$ as a discrete probability distribution over the models in our set. There are then two ways to proceed, based on how we treat model structural

M. van Oijen, *Bayesian Compendium*,
https://doi.org/10.1007/978-3-031-66085-6_11

error (or 'discrepancy'). The most principled approach is to explicitly include the discrepancy in the analysis and try to quantify it. That will be discussed in Chap. 12. Alternatively, we may make the additional assumption that one of the models in our set is correct, i.e. has no discrepancy at all. That simplifies the treatment considerably because we can then find data and use Bayes' Theorem to reduce the uncertainty about which model is correct:

$$p[f|y] \propto p[f]\, p[y|f], \tag{11.1}$$

where the model likelihood, $p[y|f]$, requires integrating out parameter uncertainty:

INTEGRATED LIKELIHOOD:

$$p[y|f] = \int p[y|\theta, f]\, p[\theta]\, d\theta. \tag{11.2}$$

We refer to $p[y|f]$ as the *integrated likelihood* for model f. Other common names are *marginal likelihood* and *model evidence*. It is, as the last equation shows, the expectation of the likelihood value under the prior probability distribution for the parameters of a model. Let's consider how we would apply these equations to model comparison. If we have two models f_1 and f_2 and no prior preference for either (i.e. $p[f_1] = p[f_2] = 1/2$), then we can use data y to calculate the posterior odds of f_1 as the ratio of the integrated likelihoods: $\frac{p[f_1|y]}{p[f_2|y]} = \frac{p[y|f_1]}{p[y|f_2]} = \frac{\int p[y|\theta_1, f_1]p[\theta_1]d\theta_1}{\int p[y|\theta_2, f_2]p[\theta_2]d\theta_2}$. That ratio of integrated likelihoods is called the *Bayes Factor* and gives us the formally correct way to use data for updating model odds (Kass & Raftery, 1995). (For incorrect, approximative ways, see Chap. 13.)

11.2 Bayesian Model Comparison (BMC)

We thus find that the key quantity in *Bayesian Model Comparison* (BMC) is each model's integrated likelihood. The fact that it is calculated by integrating over a model's whole parameter space means that it has an in-built 'Occam's razor' that favours model simplicity (MacKay, 1992a). Parameter-rich models may achieve high likelihood in some small part of their parameter space, but the part of parameter space where they perform poorly will be much larger than for lower-dimensional models. Parameter richness will thus only confer high integrated likelihood if the model's parameter distribution restricts its parameter values to the parts of parameter space where the model fits the data well. In practice, this means that complex models will only be preferred if sufficient data have been available to calibrate their parameters well. So when comparing two or more models, there is no need to penalise parameter richness in some ad-hoc way: Bayes' Theorem will take care of that automatically through the integrated likelihood. Jaynes (2003: Fig. 4.1) gives a nice example of BMC in which the relative probabilities of three hypotheses change with gradually accumulating evidence. An advance warning though: reliably estimating the integrated likelihood for a model is difficult if parameter space is high-dimensional (Kass & Raftery, 1995; Mattei, 2019). However, stable estimation methods have recently been developed (Dunstan et al., 2022; Polanska et al., 2023).

When comparing process-based models, there is the difficulty that some models may produce a different suite of output variables than other models do. To keep their integrated likelihoods comparable, we need to restrict ourselves to observations on variables that all the models can predict. Moreover, the data should obviously not already have been used for parameter calibration of any of the models. There is also the issue that some models may have been calibrated much better than others. For example, we may learn that some colleagues have produced a forest model that works very well for the tree species in their country, and we want to compare their model to our own. In that case, it is useful to have new data that can be split into a calibration set and a model comparison set, as was done by Van Oijen et al. (2013b) in their BMC of six different process-based forest models. An alternative approach is *Reversible Jump MCMC* (RJMCMC) (Green, 1995) that achieves parameter calibration and model comparison by MCMC in one go, using just one dataset, by joining the models' parameter spaces together and adding a 'model indicator parameter' whose discrete posterior probability distribution is the sought after $p[f|y]$. This works well for parameter-sparse models.

One final remark about BMC. The integrated likelihood is the expectation value of the likelihood under the prior distribution for the parameters, so its value depends not only on the data but also on the prior parameter distribution for the parameters. Now imagine that two researchers want to compare their models without knowing that they have the same model... The only difference would be their expertise concerning plausible parameter values. So they would assign different parameter priors, leading to different integrated likelihoods. The researcher with the greatest expertise would probably achieve a higher integrated likelihood than her counterpart—her parameter values will make more sense—and 'her' model would be given the highest posterior probability in the BMC. But how can that be correct if both models are the same? Well, the answer *is* correct and confirms that models without a parameter distribution are not completely defined. A model is only as good as its parameter distribution allows it to be. Models cannot be compared on their structure alone, we always evaluate 'model + parameter distribution'. You may want to make sure that different models have been calibrated on similar data in the past before comparing them (with their updated parameter distributions) against new data. Otherwise, the best calibrated model gets a spurious advantage.

11.3 Bayesian Model Averaging (BMA)

After we have carried out a BMC, we can account for model uncertainty in our predictions. We make use of the whole model ensemble by weighing each model's contribution to prediction with its posterior probability $p[f|y]$. This is referred to as *Bayesian Model Averaging* (BMA) (Kass & Raftery, 1995; Piironen & Vehtari, 2015; Van Oijen et al., 2013b). Formally:

BMA after BMC:

$$p[z = \hat{z}|y] = \sum_{i=1}^{n_f} p[f_i|y]\, p[z = \hat{z}|f_i], \tag{11.3}$$

where n_f is the number of models in our ensemble. Note that BMA does not average predictions but probabilities. For argument's sake: if we have two equally plausible models, and their predictions are $z = 10$ and $z = 20$ with complete certainty, then the BMA is not $z = 15$ but $\{p[z = 10] = 0.5;\ p[z = 20] = 0.5\}$. So predictive distributions from BMA are usually multimodal. Despite that, their predictive capacity tends to be superior to those of any single model in the ensemble (Piironen & Vehtari, 2015). But more importantly, BMA is the correct way of accounting for the uncertainty embodied in our ensemble and is therefore to be preferred over just selecting the most plausible model and discarding the others. Let's now look at an example of BMC and BMA in action.

11.4 BMC and BMA of Two Process-Based Models

We shall be comparing two versions of the expolinear process-based model (Goudriaan & Monteith, 1990). We already presented the first version, `EXPOL5`, in Chap. 8, where we calibrated its parameters by means of MCMC. `EXPOL5` is a 5-parameter model that—unusual for process-based models—can be analytically solved.

Our second model is `EXPOL6` which was also proposed by Goudriaan. It is a different analytically solvable version of the expolinear model. Just like `EXPOL5` it generates output for two variates, W and LAI. As you will expect, `EXPOL6` has 6 parameters rather than 5. The extra parameter is $LAIMAX$, the maximum value of the leaf-area index (LAI; $m^2\ m^{-2}$). So in this model, the LAI does not continue growing proportionally to biomass W, but levels off to an asymptote. Here is the R-function `EXPOL6` from which you can infer the model formula yourself:

```
EXPOL6 <- function( t=0, b=c(10,1,3,0.007,2,1) ) {
  I0  <- b[1] ; K   <- b[2] ; LAIMAX <- b[3]
  LAR <- b[4] ; LUE <- b[5] ; W0     <- b[6]
  e1  <- exp(I0*K*LAR*LUE*t) ; e2 = exp(K*LAR*W0) ; e3 = exp(-K*LAIMAX)
  W   <- W0*e3 + ( log(e2+1/e1-1)/(K*LAR) + I0*LUE*t ) * (1-e3)
  LAI <- log( (1+e1*(e2-1)) / (1+e1*(e2-1)*e3) ) / K
  return( list( "W"=W, "LAI"=LAI ) ) }
```

In Chap. 8, we calibrated EXPOL5 against a dataset on biomass W and leaf-area index LAI. To put both models on an equal footing before comparing them, we assign a similar prior to EXPOL6, extended with $LAIMAX$, and calibrate against the same data:

```
bChain <- MetropolisLogPost( f=EXPOL6, logp=logPostList, dataList=data )
```

We already showed the parameter trace plots and parameter histograms for EXPOL5 in the earlier chapter (Fig. 8.1), so we only show the MCMC results for EXPOL6 here (Fig. 11.1). Outputs of EXPOL6 for a representative sample from the posterior parameter distribution are compared to data in Fig. 11.2.

At this stage, we have defined the two expolinear models and calibrated their parameter probability distributions against the same data. Both models and their posterior parameter distributions shall now be confronted with a new dataset. This consists of data on both models' output variables, the biomass W and the leaf-area

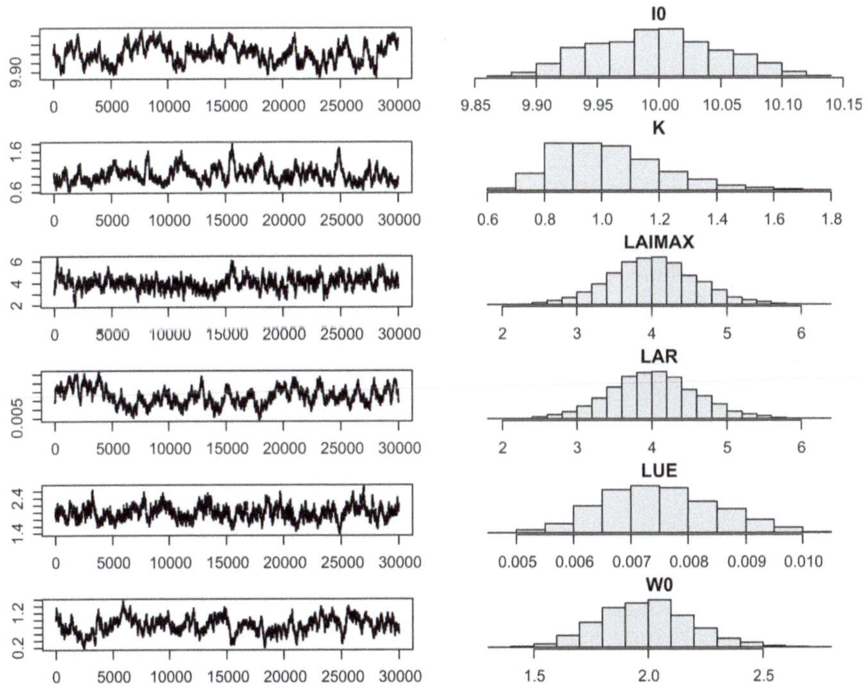

Fig. 11.1 Bayesian calibration of the 6-parameter expolinear model (EXPOL6) by means of Metropolis sampling

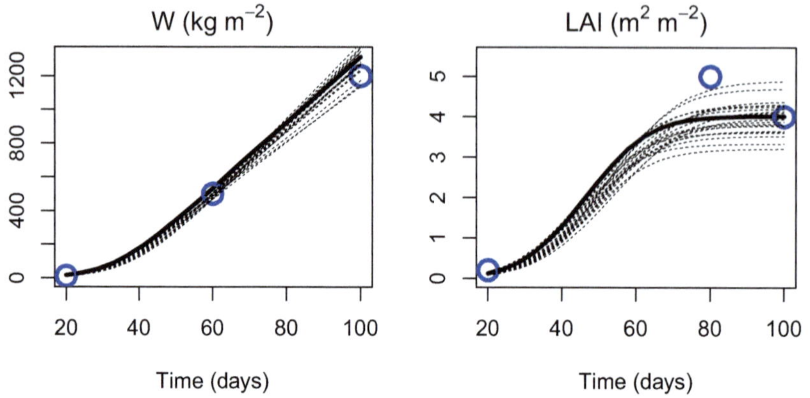

Fig. 11.2 Growth curves generated by model EXPOL6 for biomass (W) and leaf-area index (LAI). The model was run with the posterior mean parameter vector (thick line) and with 20 parameter vectors subsampled from the MCMC (thin dashed lines)

index LAI. As is typical in practice, we shall not have the same number of data points for both variables. In this case, we have fewer observations for LAI than for W.

```
t_W <- c(10,20,40,80) ; y_W <- c(5,15,150,900) ; sd_W <- c(2,4,20,100)
t_L <- c(20,80)       ; y_L <- c(0.2,4.0)      ; sd_L <- c(0.1,0.1)
data_W   <- cbind( t=t_W, y=y_W, sd=sd_W )
data_LAI <- cbind( t=t_L, y=y_L, sd=sd_L )
data_NEW <- list( "W"=data_W, "LAI"=data_LAI )
```

We first calculate the integrated likelihood EXPOL5 using the new dataset. We do this by taking a large sample of parameter vectors from the joint probability distribution for the model's five parameters, calculating the data likelihood for each parameter vector and taking the average. The following code essentially does that, although it works with log-transformed probabilities to make the calculation more numerically stable.

```
ns            <- 1e3
sample_b5     <- rmvnorm( n=ns, mean=mb5_y, sigma=Sb5_y )
sample_log_L5 <- sapply( 1:ns, function(i) {
  logLikList( sample_b5[i,], dataList=data_NEW, f.=EXPOL5 ) } )
log_IL5       <- mean(sample_log_L5) +
                 log( mean( exp(sample_log_L5-mean(sample_log_L5)) ) )
```

Note that we resampled the posterior parameter distribution from the multivariate normal with the posterior mean and covariance matrix that we found in Chap. 8. This assumes that the posterior is well represented by that normal distribution. We could instead have directly subsampled from the MCMC chain that we generated when calibrating EXPOL5 on the first dataset, and that is generally the more reliable method. We now carry out the same sampling for EXPOL6 to calculate the integrated likelihood of that model.

After that is done, and the log-transformation is reversed, we find that the two integrated likelihoods are:

- $p[\, y \mid \texttt{EXPOL5} \,] = 4.825 \times 10^{-7}$,
- $p[\, y \mid \texttt{EXPOL6} \,] = 7.702 \times 10^{-7}$.

So the posterior probabilities $p[f_i|y]$ are:

- $p[\, \texttt{EXPOL5} \mid y \,] = 0.3852$,
- $p[\, \texttt{EXPOL6} \mid y \,] = 0.6148$.

We conclude that the new dataset makes the six-parameter version of the model more plausible than the five-parameter version.

Finally, these results allow us to use BMA for prediction. Say we want to predict $W(120)$, the biomass at time 120. The proper procedure would then be to: (1) propagate parameter uncertainty for both models to gives us two predictive probability distributions for $W(120)$, (2) mix the two distributions, following the prescription of Eq. (11.3), using the two posterior model probabilities $p[f_i|y]$ that we just found.

Exercise

1. **Model precision vs. accuracy.** Say you have two models, M1 and M2. Sampling from their priors leads to normally distributed output: M1 predicts $z = 8 \pm 1$, M2 predicts $z = 10 \pm 4$. So M1's prior predictive output is the most precise. Then new data come in: $y = 10 \pm 3$. M2's output thus turns out to be more accurate. Which model has the highest posterior probability, M1 or M2?

Chapter 12
Discrepancy

We have seen that if we have a probability distribution for our model's parameters, then we can sample from that distribution to see how parameter uncertainty translates into predictive uncertainty. And if we get new data, then we can use Bayes' Theorem to update the parameter distribution and thereby reduce our predictive uncertainty. So far, so good. But a more difficult problem is that of uncertainty about model structure. We know that all models are wrong, and we discussed model diagnostics in Chap. 10. But we never know exactly how wrong our models are! In the preceding chapter on Bayesian Model Comparison (Chap. 11), we addressed the issue of model structural uncertainty by collecting an ensemble of models, rich enough that we could believe it contained the 'correct' model, followed by using data to help us quantify the relative plausibilities of the different models. But what if we are not convinced that our ensemble contains a good model, let alone a correct model? And what if we have only one model that we can work with? In such cases, we need to tackle the issue of model deficiency, or *discrepancy*, head-on.

Let's take a step back and discuss the inevitable limitations of scientific modelling. We can define *theory* as the collection of generally accepted causal relationships in a scientific discipline. In well-developed fields, such as engineering and architecture, effective models can sometimes be derived from theory alone. In other fields, such as medicine, theory is often lacking. The environmental and ecological sciences are somewhere in between. We have theory about many components of the earth system, but not enough to derive unique models from. Therefore, a wide range of modelling approaches is still in use, ranging from simple statistical models to complex process-based ones, as we see in this book. But whatever model we use, it is only an approximation to nature, so it is strictly speaking an incorrect representation of nature. To quantify this structural error, we use the term *discrepancy* (Chandler, 2013; Rougier, 2007). It is defined as the model error that remains when all parameters have been set to the best possible value. So the discrepancy is the difference between model behaviour for optimum parameter

M. van Oijen, *Bayesian Compendium*,
https://doi.org/10.1007/978-3-031-66085-6_12

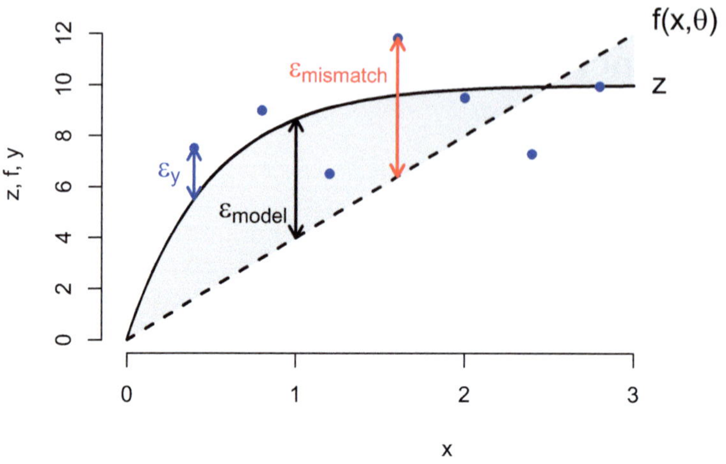

Fig. 12.1 Error types. Solid line is unknown reality z, dashed line is model output $f(x, \theta)$, blue dots are data y. Model error ϵ_{model} depends on model structure f, conditions x and model parameterisation θ. If θ is optimal, then the remaining model error is called the *discrepancy*; it is due to erroneous model structure. Because z is never fully known, we always remain uncertain about ϵ_{model} and ϵ_y

values and the true behaviour of the natural system (Fig. 12.1). Unfortunately, we never know exactly how large the discrepancy is. If we did, we would modify our model. Our uncertainty about the discrepancy complicates our attempts at Bayesian calibration (Kennedy & O'Hagan, 2001).

12.1 Treatment of Discrepancy in Single-Model Calibration

Bayesian calibration combines prior information with the likelihood, and it is in the latter term that model outputs are compared with data. There are three different ways in which the discrepancy can be treated in the likelihood: (1) ignore it, (2) lump it with data error, (3) account for data error and discrepancy separately. We now discuss the merits of the three methods.

First, we may choose not to represent the discrepancy at all, in which case our calibration is formally conditional on the model being correct. For ease of exposition, let us assume that error uncertainties are represented by zero-mean Gaussians. Then our likelihood function would read:

$$p[y|\theta] = N[y - f(x, \theta); \mu = 0, \sigma^2 = \sigma_y^2], \tag{12.1}$$

where we have the usual choice of pre-specifying the value of σ_y^2 or adding it to the parameters to be calibrated, in which case we must specify a prior distribution for it. Ignoring the discrepancy in this way is quite common; it is akin to linear regression without considering the possibility of a nonlinear function. It is a simple method that underestimates predictive uncertainty of the model because it can only propagate input uncertainty, not structural uncertainty. This method has been used repeatedly in vegetation modelling (e.g. Van Oijen et al., 2005), but we may want to move away from it.

The second approach also does not recognise the discrepancy explicitly but estimates the sum of data and model structural error, i.e. the overall mismatch between data and model outputs ($\epsilon_{\text{mismatch}}$ in Fig. 12.1). Using Gaussians again, the likelihood would then read:

$$p[y|\theta] = N[y - f(x, \theta); \mu = 0, \sigma^2 = \sigma^2_{\text{mismatch}}], \tag{12.2}$$

where we make no attempt to decompose $\sigma^2_{\text{mismatch}}$ into contributions from data error vs. discrepancy. In this method, the $\sigma^2_{\text{mismatch}}$-term cannot be pre-specified—because we cannot foresee how much model outputs will deviate from data—and must be added as a parameter to be calibrated. This method was used by, amongst others, Kavetski et al. (2006). Despite being simple, this method is also not to be advised, as it obscures the difference between errors in data and model (Rougier, 2007). It also leads to overestimation of predictive uncertainty: we do not want to predict future measurements ($f(x, \theta) + \epsilon_{\text{mismatch}} = f(x, \theta) + \epsilon_y + \epsilon_{\text{model}}$) but future true values ($f(x, \theta) + \epsilon_{\text{model}}$).

This leaves the third method, rarely used but to be recommended: explicitly distinguishing errors in data from those in model structure (ϵ_y and ϵ_{model} in Fig. 12.1):

$$p[y|\theta] = N[y - f(x, \theta); \mu = 0, \sigma^2 = \sigma^2_y + \sigma^2_{\text{model}}], \tag{12.3}$$

where we must specify a prior for σ^2_{model}. In this approach, we can account for all sources of model predictive uncertainty ($p[x]$, $p[\theta|y]$, $p[\epsilon_{\text{model}}]$). Although this method is in principle superior to any other, it may be difficult to specify the prior for the discrepancy because simple Gaussians (or Gaussian Processes, see Chap. 15) are unlikely to be adequate. However, the method has been sketched out in ground-breaking papers (Kennedy & O'Hagan, 2001; Rougier, 2007) and is being developed further (Bradley et al., 2022; Cameron et al., 2022; Spitieris & Steinsland, 2023). Brynjarsdóttir and O'Hagan (2014) provide a very clear discussion of the need for discrepancy modelling and its difficulties.

12.2 Treatment of Discrepancy in Model Ensembles

So far, we have discussed two ways of accounting for the fact that every single model is wrong: ensemble modelling in Chap. 11 and discrepancy quantification in this chapter. Let's now attempt to combine both methods. We start with the conceptual diagram shown in Fig. 12.2.

The figure shows the behaviour of a system in reality as well as the behaviour of models. The figure is a map, with every map location representing a different trajectory of the system state, i.e. a time series for the multivariate system state of interest. So despite the figure being two-dimensional, every point represents an ultra-high dimensional quantity. On the left we see reality, but the models are on the right, implying that all of them are wrong. Each model's distance from reality measures its discrepancy. The models are clustered, so to some degree they share a common bias. That means we can decompose each model's discrepancy as the sum of a shared bias and a model-specific deviation from the centre of the cluster. The figure is a simplification of a scheme proposed by Chandler (2013) for climate model ensembles, based on the realisation that there is always some convergence in model development, leading to structural errors that are shared by all models in the ensemble. Chandler and colleagues suggest that we try to quantify the two components of discrepancy (ensemble bias and model-specific deviation) separately (Spence et al., 2018). A similar decomposition of discrepancy was proposed by Goldstein (2011), who used the terms *internal* and *external structural uncertainty*. Goldstein provided suggestions for the kinds of model structural and parametric sensitivity analysis that a modeller could perform to quantify internal uncertainty, and the more informal expert elicitation needed to estimate external uncertainty (Goldstein & Rougier, 2009). Overall, this approach seems attractive. It accounts, in principle, for model structural uncertainty in the most comprehensive and realistic way. However, it remains hard to make all steps fully operational. If we want to estimate the scheme components in a Bayesian way, then we need to assign prior

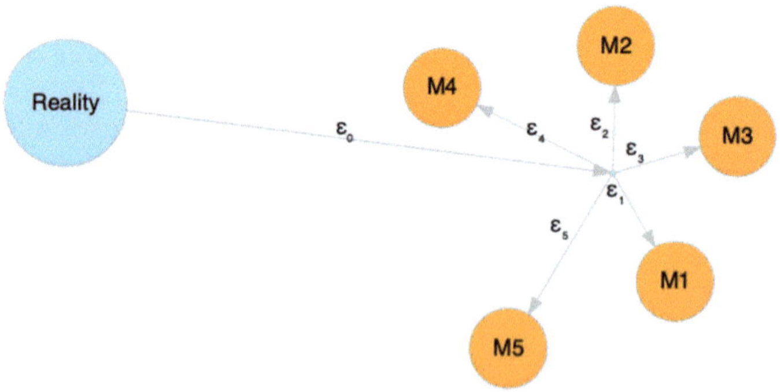

Fig. 12.2 Conceptual diagram for model discrepancy, showing model-specific biases and an overall bias shared by the whole cluster. Simplified after (Chandler, 2013)

probability distributions to all biases (for ensemble mean and for individual models), but we generally have no heuristics telling us how to do so. At the very least, Fig. 12.2 can keep us on our toes by reminding us that convergence of models does not imply convergence to the truth!

Chapter 13
Approximations to Bayes

Bayes' Theorem tells us how to learn from data. We just need to assign our prior probability distribution for parameters or models and formulate the data likelihood function. The posterior distribution is then fully determined, and it encapsulates everything of interest. With the posterior in hand, we can make predictions with proper uncertainty quantification, we can carry out risk analysis and provide decision support. So the basic ideas are extremely simple and powerful. But we have also seen that assigning a prior and likelihood is not always easy, and deriving a sample from the posterior distribution may require computationally demanding methods such as MCMC. So people keep searching for shortcuts where the Bayesian analysis can be made faster albeit perhaps a little bit less informative and accurate. Martin et al. (2023) provide a good overview of these various attempts. Some of the approximation methods date from times when computers were slow and application of sampling-based Bayesian analysis was inevitably time-consuming. But the problem with computational demand has not gone away. The advent of complex computer simulators, such as global climate models (GCM) and dynamic vegetation models (DGVM), is keeping computational efficiency high on the agenda. There will always remain an important role in computational statistics for approximation methods. But I do believe that they are over-used. There is no better way of assessing your own understanding of a system than thinking carefully about all the system's parameters and about all measurements and their uncertainties. Writing down a prior and likelihood function is a learning experience in itself. And the posterior gives you more information than any of the outcomes from approximation can do. Another reason to prefer the real Bayesian deal over approximation methods is that Bayes' Theorem makes no assumptions about how many data you have (you don't need infinitely many) or about your parameter uncertainties (they need not be uniform or Gaussian) or your model (it need not be linear in any way). Bayes gives you the freedom to express exactly what you know and don't know. Bayes' Theorem can handle it.

M. van Oijen, *Bayesian Compendium*,
https://doi.org/10.1007/978-3-031-66085-6_13

This book focuses on explaining the basic applications of Bayesian analysis, so we shall not discuss any of the approximation methods at length. But here is very briefly a list of some of the most important ones with pointers to the literature.

13.1 Approximations to Bayesian Calibration

Methods for approximating and speeding up Bayesian calibration of model parameters include:

- *Approximate Bayesian Computation* (ABC). In ABC, you do not formulate a likelihood function. Instead you evaluate the likelihood of a parameter vector $p[y|\theta]$ by generating multiple different 'observations' from which you calculate summary statistics that you compare with the corresponding summary statistics of the real data to feed into an accept–reject rule. This is more suitable for stochastic models or chaotic models with initial-state uncertainty than for deterministic simulators. For more, see: Lintusaari et al. (2017).
- *Linear Bayes* (LB). In LB, the whole idea of formulating joint probability distributions is given up. The two central equations of LB, for adjusting expectation and variance, resemble the Kalman Filter equations. For more, see: Goldstein (2015). Nott et al. (2014) interpret ABC within the framework of LB.
- *Laplace Approximation* (LA). LA assumes that the posterior distribution is a multivariate Gaussian. Bayesian calibration using this method first finds the MAP and treats that as the mean of our Gaussian. Then it inspects the curvature of the density around that parameter vector to estimate the covariance matrix. In technical terms, the covariance matrix is taken to be the inverse of the negative Hessian matrix (matrix of second-order partial derivatives) of the posterior log density around the MAP. This is a good approximation when the shape of the posterior distribution is close to Gaussian near its mode. An early example of the method (but not using the name LA) was MacKay (1992b).
- *Integrated Nested Laplace Approximation* (INLA). INLA makes assumptions of linearity and uses Laplace Approximations of probability distributions to arrive at marginal posterior distributions only, i.e. the distributions for individual parameters. No posterior parameter correlations are quantified. INLA is designed for hierarchical models and treats parameters and hyperparameters differently (Margossian et al., 2020). It has been adopted widely in the spatial modelling community. For more information, see: Blangiardo et al. (2013), Opitz (2017), Bakka et al. (2018), and Simpson et al. (2023).
- *Variational Bayes* (VB). VB has some similarity to LA in that it aims to approximate the posterior with a well-behaved probability distribution, often a Gaussian, but it does not start from the MAP. Instead it uses an iterative algorithm to find those hyperparameter values of the approximating probability distribution that minimise the overall distance (measured by the so-called *Kullback-Leibler divergence*) between the approximating distribution and the

true posterior. It is frequently used in machine learning (Barber & Bishop, 1998). Clear introductions to VB for practitioners are given by Bishop (2006) and Tran et al. (2021); a more technical exposition is Blei et al. (2017).

13.2 Approximations to Bayesian Model Comparison

Bayes' Theorem implies that the *integrated likelihood* (IL) is the main criterion for model comparison (see Chap. 11). However, the IL is difficult to quantify reliably, especially for high-dimensional parameter spaces. Variational Bayes (see above) provides a lower bound to the IL as part of its distribution optimisation. Alternatively, various so-called *information criteria* have been proposed for model evaluation and selection, each of which can be seen as an approximation to the IL. One of the oldest and still most widely used is the *Akaike Information Criterion* (AIC) which is defined as follows:

$$AIC = -2\,log(\,p[y|\theta^{MLE}]) + 2\,n_p,$$

where θ^{MLE} is the maximum likelihood parameter estimate and n_p is the number of parameters in the model. Models with low AIC are preferred. But you notice the absence of a prior distribution in this definition and instead there is the penalty proportional to the number of parameters. The AIC only ranks models reliably when there are many data (so the prior is of little importance) and when the model is fairly linear—but those are conditions for which the IL can be easily calculated as well. Similar reservations can be expressed against the whole zoo of information criteria which nowadays include the WAIC, BIC, DIC and many more. For more information, see: Burnham and Anderson (2002), Bolker (2008), Hojsgaard et al. (2012) and Gelman et al. (2014).

Chapter 14
Thirteen Ways to Fit a Straight Line

14.1 Hidden Equivalences

Fitting a straight line through data can be done in many ways that may seem different at first, but after closer inspection prove to be based on the same mathematics. In this chapter, we shall fit a line to data in 13 different ways and compare the resulting parameter estimates. So our goal is to estimate the intercept and the slope of the line. This is our line-up of algorithms:

1. Ordinary Least Squares (OLS) = R-function 'lm'
2. Weighted Least Squares (WLS)
3. Generalised Least Squares (GLS)
4. Bayesian calibration analytically (Lindley & Smith, 1972)
5. Kalman Filtering (KF)
6. Gaussian Process (GP)
7. Conditional multivariate Gaussian
8. Graphical model (GM)
9. Machine learning with a Neural Network (NN)
10. Accept–Reject Monte Carlo sampling (A-R)
11. Metropolis MCMC
12. Gibbs Sampling MCMC
13. R-package JAGS (also MCMC)

We are going to assume that we have the same measurement uncertainty for each data point, and that our prior uncertainty about the two parameters is very large. It then turns out that the first eight methods will give exactly the same results for the parameter estimates, because they are all based on the same underlying mathematics for Gaussian linear models. The other algorithms will give the same results, but with minor deviations because of sampling variability.

The goal of this exercise is threefold. First, I like to show that ostensibly 'classical' methods, such as the least squares estimations, are the same as Bayesian

M. van Oijen, *Bayesian Compendium*,
https://doi.org/10.1007/978-3-031-66085-6_14

calibration given certain assumptions for the prior and likelihood. Secondly, I hope that the exercise, by convincing you of the strong unity underlying most statistical approaches, will reassure you that there is much less to learn than confusing terminology might suggest. Finally, this chapter will be a gentle first introduction to some methods that we will study in greater detail in later chapters.

14.2 Our Data

We shall mainly be using the following data in our examples.

```
x <- c(10,20,30) ; y <- c(6.09,8.81,10.66) ; ny <- length(y)
```

The vector x contains three values of our scalar predictor variable (which could for example be time or distance or temperature), and the vector y contains the corresponding observations for the dependent variable.

14.3 The Normal Equations for Ordinary Least Squares (OLS)

Say we want to fit a straight line to our data. Classical statistics has provided us with simple equations that we can use for this purpose, called the Normal Equations:

$$\begin{aligned} &\underline{\text{NORMAL EQUATIONS}} \text{ for } y = \beta_1 + \beta_2 x : \\ &\text{Estimate of slope: } \hat{\beta}_2 = \frac{\text{Cov}(x, y)}{\text{Var}(x)} \\ &\text{Estimate of intercept: } \hat{\beta}_1 = \bar{y} - \hat{\beta}_2 \bar{x} \\ &\text{Correlation coefficient: } \rho = \frac{\text{Cov}(x, y)}{\sigma_x \sigma_y}, \end{aligned} \tag{14.1}$$

where the hat-symbol ^ stands for 'estimated value of' and the overbar-symbol ¯ for 'mean of'. Using these equations is referred to as Ordinary Least Squares regression, or OLS for short. Let's program the equations in R.

```
b2 <- cov(x,y) / var(x)
b1 <- mean(y) - b2 * mean(x)
```

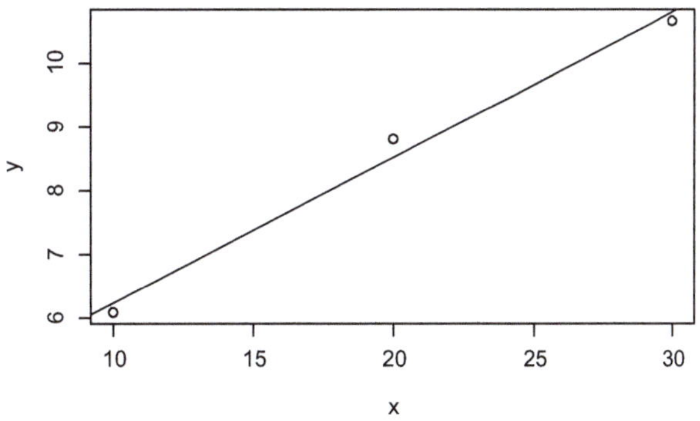

Fig. 14.1 Linear regression on three data points

Running these two lines of code produces the following results.

$$\frac{\text{ORDINARY LEAST SQUARES (OLS):}}{\hat{\beta_1} = 3.95; \quad \hat{\beta_2} = 0.2285.} \tag{14.2}$$

We see that the value of the intercept is estimated as 3.95, and the value of the slope as 0.2285. The equations are simple and easy to implement, but we could have chosen to use the R-function 'lm' which also carries out OLS using the Normal Equations and therefore gives the same results:

```
  lm.y <- lm(y~x) ; print(coef(lm.y))
> (Intercept)          x
>      3.9500     0.2285
```

Figure 14.1 shows the data and the line that we have fit.

14.3.1 Uncertainty Quantification

OLS is not Bayesian. In Bayesian modelling, we need to state how we treat uncertainty about data error and what priors we assign to parameters. In non-Bayesian methods such as OLS using R's `lm`-function, those parts of the problem are swept under the carpet: data uncertainty is estimated from the residuals when fitting the model, and priors are assumed to be uniform on the domain from minus to plus infinity.

So if we use the Normal Equations or '`lm`', we do not have to specify our uncertainty about errors in the measurements y. OLS is an estimation method which does not make distributional assumptions. Therefore, without further information, this method cannot quantify uncertainty about the estimated parameter values. But if we assume that: (1) our model and parameter estimates are correct, (2) measurement errors are all drawn independently from the same zero-centred Gaussian distribution, then we can estimate the variance of that error distribution from the properties of the residuals. And then we can quantify what that measurement uncertainty implies for uncertainty about the intercept and slope. In fact, with these assumptions there are simple analytical formulas for the posterior variances of slope and intercept as well as their covariance. But instead of giving these, we immediately move to more general and powerful equations that do not require measurement errors to be independent and identically distributed. The solution for OLS is then just a special case.

14.4 Generalised Least Squares (GLS)

We now fit the line to our data using the equations for Generalised Least Squares (GLS) estimation. This requires us to specify uncertainty about measurement errors in the form of a covariance matrix Σ_y, with the variances on the diagonal and the error-covariances in the other positions. Strictly speaking, GLS makes no distributional assumptions, but in regression practice Σ_y is commonly assumed to be the covariance matrix of a multivariate Gaussian distribution. GLS allows us to have more than one independent predictor variable, so instead of just x, we could have $x^{(1)}, .., x^{(n_x)}$. We collect the predictor variables in the so-called *design matrix* $\mathbf{X}$, which is a matrix of magnitude $n_y \times (n_x + 1)$ with a column of ones on the left followed by n_x different columns for $x^{(1)}, .., x^{(n_x)}$. We also collect the parameters in one vector $\beta = (intercept, slope^{(1)}, .., slope^{(n_x)})$.

The GLS equations can then be written very compactly as:

$$\begin{aligned} \text{Model: } & y = \mathbf{X}\beta + \epsilon \quad \text{with } \epsilon \sim N[0, \Sigma_y] \\ \text{Estimate of covariance matrix: } & \widehat{\Sigma_\beta} = (\mathbf{X}^\top \Sigma_y^{-1} \mathbf{X})^{-1} \\ \text{Estimate of parameter vector: } & \hat{\beta} = \widehat{\Sigma_\beta} \mathbf{X}^\top \Sigma_y^{-1} y. \end{aligned} \tag{14.3}$$

In this chapter, we only have one predictor variable, so $\mathbf{X}$ will be a $n_y \times 2$ matrix, and β will just be $(intercept, slope)$. Let's now implement the GLS equations in R. As mentioned, that requires us to specify the covariance matrix for measurement error Σ_y, which we initially shall choose to be a diagonal matrix with the same value everywhere on the diagonal:

$$\Sigma_y = \begin{bmatrix} 3 & 0 & 0 \\ 0 & 3 & 0 \\ 0 & 0 & 3 \end{bmatrix} \tag{14.4}$$

We now implement the GLS equations in R.

```
X    <- cbind( 1, x )                            # Design matrix
Sb_y <- solve(   t(X) %*% solve(Sy) %*% X ) # GLS variance
mb_y <- Sb_y %*% t(X) %*% solve(Sy) %*% y    # GLS estimator
```

When running the code, we get the following results.

GENERALISED LEAST SQUARES (GLS):

$$\hat{\beta} = \begin{bmatrix} 3.95 \\ 0.2285 \end{bmatrix}; \quad \widehat{\Sigma_\beta} = \begin{bmatrix} 7 & -0.3 \\ -0.3 & 0.015 \end{bmatrix}. \tag{14.5}$$

So we see that we get the same estimates for the slope and the intercept as with OLS, but we also get a covariance matrix for the parameters, representing our uncertainty about those estimates.

14.4.1 From GLS to WLS and OLS

To retrieve the OLS equations from GLS, the covariance matrix for measurement error must have the same value everywhere on the diagonal, and zeros everywhere else. Such a diagonal matrix can be written as a scalar σ_y^2 times a $n_y \times n_y$ identity matrix: $\Sigma_y = \sigma_y^2\,\mathbf{I}$. And that is in fact what we did in our example, which therefore was both GLS and OLS.

The example of GLS that we gave here is also an example of Weighted Least Squares (WLS) regression which is intermediate between OLS and GLS. WLS allows the different data points to have different error variances (which will act as inverse weights), but it requires that the data uncertainties are not correlated. So the measurement error covariance matrix in WLS must be diagonal, as was the case for our Σ_y. In short:

- OLS: Measurement errors uncorrelated and equal variance $\rightarrow \Sigma_y = \sigma_y^2\,\mathbf{I}$,
- WLS: Measurement errors uncorrelated and unequal variance $\rightarrow \Sigma_y =$ diagonal,
- GLS: Measurement errors correlated and unequal variance $\rightarrow \Sigma_y =$ symmetric positive definite.

Why do we pay so much attention to these classical equations for linear regression in a book about Bayesian methods? Well, the GLS equations can be interpreted as Bayesian linear regression using Gaussian priors and likelihoods. GLS is the special case where the Gaussian prior has such a large variance that it effectively becomes a uniform prior.

14.5 The Lindley and Smith (LS72) Equations

The GLS can be generalised to a fully Bayesian approach with an informative prior, i.e. with prior parameter variances that do not need to be infinite. Lindley and Smith (1972) (hereafter LS72) proved this by deriving equations for the posterior mean and covariance matrix of the linear model with Gaussian distributions for parameter prior and measurement error uncertainty. Here are their equations:

$$\begin{aligned}\text{Model: } y &= \mathbf{X}\beta + \epsilon \quad \text{with } \beta \sim N[\mu_\beta, \Sigma_\beta], \epsilon \sim N[0, \Sigma_y] \\ \text{Posterior covariance matrix: } \Sigma_{\beta|y} &= (\Sigma_\beta^{-1} + \mathbf{X}^\top \Sigma_y^{-1}\mathbf{X})^{-1} \\ \text{Posterior mean: } \mu_{\beta|y} &= \Sigma_{\beta|y}(\Sigma_\beta^{-1}\mu_\beta + \mathbf{X}^\top \Sigma_y^{-1} y).\end{aligned} \tag{14.6}$$

And this is the very short R-code that implements these equations:

```
Sb_y_LS72 <- solve( solve(Sb) + t(X) %*% solve(Sy) %*% X ) ;
mb_y_LS72 <- Sb_y_LS72 %*% (solve(Sb) %*% mb + t(X) %*% solve(Sy) %*% y)
```

To apply the equations to our standard example of this chapter, we must now specify a prior for the regression parameters. Let's assign the following zero-mean bivariate Gaussian distribution:

$$\begin{aligned}&\underline{\text{PRIOR FOR } \beta\text{:}} \\ &p[\beta] = N[\mu_\beta, \Sigma_\beta], \text{ where} \\ &\mu_\beta = \begin{bmatrix} 0 \\ 0 \end{bmatrix}; \quad \Sigma_\beta = \begin{bmatrix} 10000 & 0 \\ 0 & 10000 \end{bmatrix}.\end{aligned} \tag{14.7}$$

When you now run the LS72-code with this prior, you get the same results as we found with GLS (Eq. (14.5)). The results are the same despite the introduction of the prior probability distribution for the two model parameters because we set the prior variances in Eq. (14.7) so high that only the likelihood was informative, which is the underlying assumption of GLS.

The GLS equations are a special case of the LS72 equations with the prior parameter variance going to infinity and Σ_β^{-1} thus going to zero. You can easily see this by removing the terms involving Σ_β^{-1} from the LS72 equations (i.e. `solve(Sb)` in the R-code).

14.6 Kalman Filter

Now let's look at equations that at first look completely different from the equations shown above but are actually mathematically the same: the Kalman Filter equations (Soyer, 2018).

The Kalman Filter (KF) is commonly used as a procedure for updating our estimate for the state of a dynamical system, each time when new data come in (and that is how we introduce it in Chap. 22). The KF assumes that (1) state change follows a linear function of the state itself, with known parameters, and (2) state uncertainty can be represented by a Gaussian distribution. It is highly unusual, and really not very practical, to use the KF for linear regression, i.e. for parameter estimation rather than for state estimation. But we shall do so anyway to show that, under the hood, the KF uses the same mathematics as the GLS equations.

The trick to using KF for linear regression is to realise that there is no real difference between the state of a system at a specific time and a parameter vector. Both are uncertain quantities that we want to learn about. The fact that the state of a dynamical system can be written as a function of the state that it had in the recent past, does not alter that fact—it just helps us constrain state uncertainty. So the use of KF for linear regression requires us to change our perspective: we treat the parameter vector β of the linear model (the slope and intercept) as the 'state' to be estimated with KF, and the sequence of values for the independent variable x as the known 'parameter' vector. The data y will as usual provide the information we need to carry out the estimation.

KF is a fully Bayesian procedure. As for GLS, KF requires us to begin with specifying Gaussian probability distributions for measurement errors, but it adds to that the requirement to specify a Gaussian prior for what we want to estimate (normally in KF the system state but here the parameter vector β), and we use the same prior that we defined above for the Lindley and Smith equations.

So at this stage, we have the data y with their uncertainties and the prior $p[\beta]$. The final ingredient we need for KF is the linear model itself, which in KF-parlance is referred to as the *observational operator matrix*, and is typically denoted as $\mathbf{H}$. But this matrix is in fact exactly the same as the design matrix that we already defined earlier, so we denote it as $\mathbf{X}$. The idea behind the KF-terminology is that the product of the observational operator with the state-vector plus measurement error predicts the observations y. So here $\mathbf{X}$ is again a 3 $\times$2 matrix with all ones in the first column and the values of x in column 2. That completes the specification of our linear model, and KF-theory now tells us that the solution for $p[\beta|y]$ is as follows:

$$\begin{aligned}
\text{'Kalman Gain':} \quad & \mathbf{K} = \Sigma_\beta \mathbf{X}^\top (\mathbf{X}\Sigma_\beta \mathbf{X}^\top + \Sigma_y)^{-1} \\
\text{Posterior mean:} \quad & \mu_{\beta|y} = \mu_\beta + \mathbf{K}(y - \mathbf{X}\mu_\beta) \\
\text{Posterior covariance matrix:} \quad & \Sigma_{\beta|y} = (\mathbf{I} - \mathbf{K}\mathbf{X})\Sigma_\beta
\end{aligned} \tag{14.8}$$

where $\mathbf{X}^\top$ is the transpose of $\mathbf{X}$ (rows and columns interchanged) and $\mathbf{I}$ is the identity matrix (ones on the diagonal, zeros everywhere else). We implement this in R:

```
KalmanGain <- function(B,H,R){ B %*% t(H) %*% solve(H %*% B %*% t(H) + R) }
K          <- KalmanGain( Sb, X, Sy )
mb_y       <- mb + K %*% ( y - X %*% mb )
Sb_y       <- ( diag(nb) - K %*% X ) %*% Sb
```

When you run this code, you find that the estimates for β that we find with KF are the same as with GLS and LS72. The fact that KF for linear regression gives the same parameter estimates and posterior covariance matrix shows that the methods are mathematically identical—at least if the parameter prior that we specify in KF is uninformative, which was the choice we made by assigning very large variances in our Σ_β.

14.7 Gaussian Process (GP)

We shall encounter Gaussian Processes (GP) at various places in this book, with the most detailed introduction in Chap. 15 and the most advanced application in Chap. 23. GPs are not just multivariate but 'infinite-variate' Gaussian distributions that come with a deterministic mean function for trend modelling plus a covariance function for correlations between any two of the infinite variates. That short, cryptic description will make more sense after having read the first part of Chap. 15, but for now it may be enough to state how a GP can be used for the linear regression problem of this chapter. If the correlation length ϕ for a GP is very small, close to zero, then even points that are close to each other behave independently. In other words, the trend part of the GP then dominates, and the role of the stochastic process component becomes negligible apart from quantifying the variance for data error uncertainty. Because the data points are then effectively independent conditional on the regression parameters, this approach of 'GP with $\phi \to 0$' is equivalent to linear regression if the GP trend is modelled as linear. So you could use the equations for GP (or a package for GP such as `geoR`) for linear regression if you wanted to, and we shall discuss that further in Chap. 15 (see Fig. 15.1). But this use of GP for linear regression is of course, like the use of KF discussed earlier, awkward and unpractical. In standard *Gaussian process regression* (GPR), the mean function is sometimes chosen to equal zero but never the correlation length.

14.8 Conditional Multivariate Gaussian

In the fully Gaussian world, Bayesian estimation is just a matter of manipulating mean vectors and covariance matrices. Let $z = (\beta, y)$ represent the combination of parameters and observations (that we have not seen yet). We represent our prior uncertainty about the parameters and the data in the Gaussian distribution with mean vector μ_z and covariance matrix Σ_z. What will be our posterior distribution for β once we get observational values for y?

We shall now define an R-function that calculates the mean and covariance matrix of the posterior conditional multivariate Gaussian distribution $p[\beta|y]$. You will meet the function again in Chap. 16, where its role in Gaussian Bayesian Networks will be explained. The function takes as arguments μ_z, Σ_z and a vector with observed true values y and produces as output the list $(\mu_{\beta|y}, \Sigma_{\beta|y})$.

```
GaussCond <- function( mz, Sz, y ) {
  i <- 1 : ( length(mz) - length(y) )
  m  <- mz[i]    + Sz[i,-i] %*% solve(Sz[-i,-i]) %*% (y-mz[-i])
  S  <- Sz[i,i]  - Sz[i,-i] %*% solve(Sz[-i,-i]) %*% Sz[-i,i]
  return( list( m=m, S=S ) ) }
```

Now we use this function for linear regression. But how do we determine the covariance matrix for parameters and unobserved data? In this case of linear regression with just three data points, it is not too much work to create this 5×5 matrix. The top left 2×2 will just be the prior covariance matrix for the parameters Σ_β. So the first two entries on the diagonal are the prior variances for the parameters which we denote as V_{β_1} and V_{β_2}. The three other entries on the diagonal are the prior *unconditional* variances for the as yet unobserved data $V_{y_i}^{\text{unc}}$, i=1..3. These variances will be large because they are not conditioned on β. Under our linear model, the unconditional observation variances are calculated as $V_{y_i}^{\text{unc}} = V_{\beta_1} + x_i^2 V_{\beta_2} + V_{y_i}$. Covariances between observations are calculated as $\text{Cov}_{y_i,y_j}^{\text{unc}} = V_{\beta_1} + x_i x_j V_{\beta_2}$. That leaves the covariances between parameters and observations, which we find as $\text{Cov}_{\beta_1,y_i}^{\text{unc}} = V_{\beta_1}$ and $\text{Cov}_{\beta_2,y_i}^{\text{unc}} = x_i V_{\beta_2}$. This all looks complicated, but it can be implemented in the following two lines of R-code.

```
mz <- c( mb, X %*% mb )
Sz <- rbind( cbind( Sb          ,   Vb*t(X)               ),
             cbind( t(Vb*t(X)), t(Vb*t(X))%*%t(X)+Sy ) )
```

The prior mean and covariance matrix are then found to be:

PRIOR MULTIVARIATE GAUSSIAN OF PARAMETERS AND OBSERVATIONS

$$\mu_z = \begin{bmatrix} 0 \\ 0 \\ 0 \\ 0 \\ 0 \end{bmatrix} ; \quad \Sigma_z = \begin{bmatrix} 10000 & 0 & 10000 & 10000 & 10000 \\ 0 & 10000 & 1e+05 & 2e+05 & 3e+05 \\ 10000 & 1e+05 & 1010003 & 2010000 & 3010000 \\ 10000 & 2e+05 & 2010000 & 4010003 & 6010000 \\ 10000 & 3e+05 & 3010000 & 6010000 & 9010003 \end{bmatrix} . \tag{14.9}$$

All we have to do now to update this prior multivariate Gaussian—and find the Bayesian posterior—is condition the distribution on the observations. We do this by means of our previously defined R-function GaussCond.

```
Post <- GaussCond( mz, Sz, y ) ; mb_y <- Post$m ; Sb_y <- Post$S
```

And this gives the same results as GLS, LS72 and KF, as you can verify for yourself.

14.9 Graphical Modelling (GM)

The complicated derivation of the joint covariance matrix for parameters and observations of the preceding section can be much simplified by the use of Graphical Modelling. This technique will be introduced more fully in Chap. 16. But the basic idea is simple: every GM represents a joint probability distribution as a graph with arrows between nodes that show how the joint distribution can best be decomposed into a number of simpler conditional probability distributions. In the present case of our regression problem, we define a network with five nodes: the two parameters of the straight-line model and the three observations, and we show the graph for the network in Fig. 14.2.

The theory of GM has provided us with elegant algorithms for deriving the desired covariance matrix from that graph structure (Shachter & Kenley, 1989). We shall show those algorithms, including R-code, in Chap. 16, but for now just state that the GM approach leads to the exact same covariance matrix as we cumbersomely derived in the preceding section. It therefore also leads to the same posterior distribution for the parameters in our regression problem as we have seen many times before in this chapter.

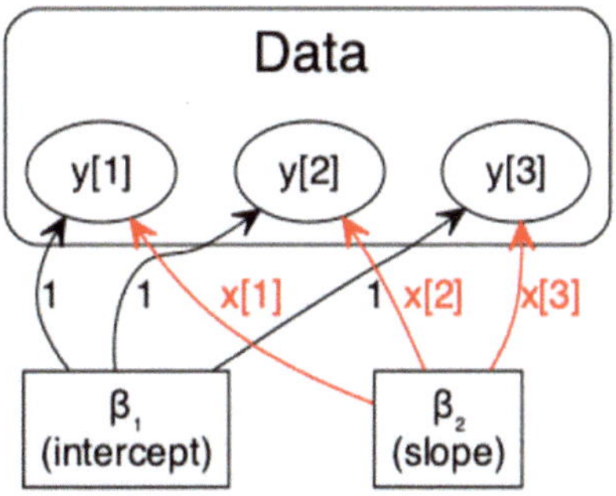

Fig. 14.2 Directed Acyclic Graph (DAG) for the linear regression graphical model

14.10 Machine Learning with a Neural Network

We shall discuss machine learning, with a focus on neural networks, in Chap. 21, but here is an ultra-short introduction. A neural network is a function that takes univariate or multivariate input, feeds that through layers of nodes, ultimately arriving at a layer with one or more output nodes. The layers between input and output are the so-called 'hidden layers'. Each node is itself a simple function, typically a weighted sum of the nodes connected to it from the preceding layer, modified by an additive bias and a usually nonlinear transformation. The simplest neural network would have one hidden layer, and just a single node in each layer, as depicted in Fig. 14.3. It would also leave out the nonlinear transformation. That network cannot do much, but it can still do linear regression which of course is nothing more than applying a weight and a bias to input. Bayesian calibration of neural networks is rarely carried out—usually machine learning amounts to parameter optimisation without uncertainty quantification. But we shall give examples of Bayesian calibration of neural networks in Chap. 21.

14.11 Accept–Reject Sampling

We showed A-R in action in Chap. 5 (see Fig. 5.1), and it can of course be used for the linear regression problem of this chapter as well. All we need to do is generate a sample from our prior distribution and use the likelihood function for pruning. We leave the implementation of A-R for this problem as an exercise for the reader, although the method is so inefficient that it may be difficult to get it to work without

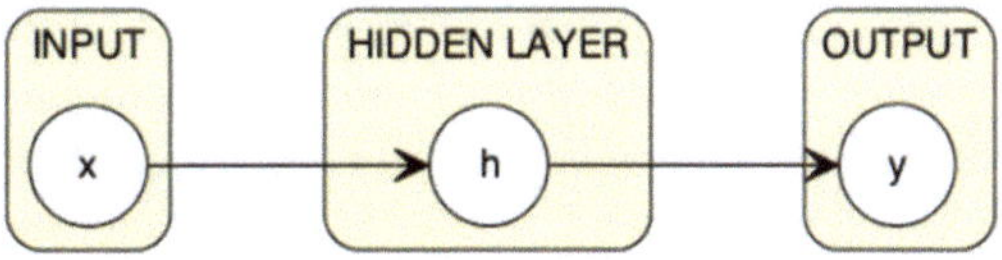

Fig. 14.3 A minimal neural network with a single hidden layer

needing an unfeasibly large sample. Rather than sampling from the full unbounded prior parameter space that we have here, you may want to limit yourself to a small part of parameter space from the beginning.

14.12 MCMC with the Metropolis Algorithm

We already showed linear regression by means of MCMC with the Metropolis algorithm in Chap. 7. You can get very close to the results of all the methods shown above by using a chain length of 10^5.

14.13 MCMC with Gibbs Sampling

We shall now fit the straight line by means of another MCMC algorithm, *Gibbs sampling*. But before we show how this can be implemented, we define an R-function `GaussMult` that will come in handy when writing code for Gibbs. `GaussMult` calculates the mean and variance of a Gaussian that is formed by multiplying (and then renormalising) n different univariate Gaussians, each with their own mean and variance.

```
GaussMult <- function( m=rep(0,2), V=rep(1,2) ) {
  n       <- length(m)
  weights <- sapply( 1:n, function(i){ prod(V[-i]) } )
  m_Mult  <- sum( m * weights ) / sum( weights )
  V_Mult  <- prod( V ) / sum( weights )
  return( list( m=m_Mult, V=V_Mult ) ) }
```

In the Gibbs sampling, we shall be using that function to calculate the product of the marginal prior and likelihoods (one for each data point) for individual parameters. We shall also be using different ways of writing the likelihood function than we are used to so far. You can verify yourself that the three lines of R-code for L below give exactly the same value.

```
i  <- 1 ; sy=sqrt(Vy) ; b1 <- 0 ; b2 <- 0
L <- dnorm( b1 + b2*x[i] - y[i], mean= 0             , sd=sy[i]        )
L <- dnorm(      b2            , mean=(y[i] - b1)/x[i], sd=sy[i]/x[i] ) / x[i]
L <- dnorm( b1                 , mean= y[i] - b2 *x[i], sd=sy[i]       )
```

We shall use the last two lines in the R-code for finding analytical solutions for $p[b_1|b_2, y]$ and $p[b_2|b_1, y]$, i.e. the conditional posterior probability distributions for one parameter given all data and the other parameter. That is the basis of Gibbs sampling: taking each parameter one at a time. Our whole Gibbs algorithm is the following:

```
ni <- 1e3 ; bChain <- matrix(NA, nrow=ni, ncol=2)
b1 <- mb[1] ; b2 <- mb[2] ;  bChain[1,] <- c( b1, b2 )
for (i in (2 : ni)) {
  mb1_L <- y - b2*x ; Vb1_L <- diag( Sy )
  Post1 <- GaussMult( m=c(mb[1],mb1_L), V=c(Vb[1],Vb1_L) )
  mb1   <- Post1$m ; Vb1 <- Post1$V
  b1    <- rnorm( 1, mean=mb1, sd=sqrt(Vb1) )
  mb2_L <- ( y - b1 ) / x ; Vb2_L <- diag( Sy/(x^2) )
  Post2 <- GaussMult( m=c(mb[2],mb2_L), V=c(Vb[2],Vb2_L) )
  mb2   <- Post2$m ; Vb2 <- Post2$V
  b2    <- rnorm( 1, mean=mb2, sd=sqrt(Vb2) )
  bChain[i,] <- c( b1, b2 ) }
```

Gibbs sampling cannot always be used: it requires that we are able to write down the formula for the conditional posterior probability for one parameter given the values of all other parameters and the data. In linear regression, Gibbs can be used because we know how to write down those equations as you can see in the R-code. Another downside of Gibbs is that it considers one parameter at a time and not the whole parameter vector as in our Metropolis implementation. So you might expect Gibbs sampling to take longer than Metropolis. But that is often not the case because Gibbs has the major advantage that every proposal is accepted. For our example in this chapter, Gibbs is better than Metropolis. The following results were produced with Gibbs using a chain length of just 1000:

GIBBS SAMPLING:

$$\mu_{\beta|y} = \begin{bmatrix} 4.39 \\ 0.207 \end{bmatrix}; \quad \Sigma_{\beta|y} = \begin{bmatrix} 6.977 & -0.294 \\ -0.294 & 0.015 \end{bmatrix}. \tag{14.10}$$

Figure 14.4 shows the frequency distributions for the two parameters generated by the Gibbs sampling.

14.14 MCMC with Gibbs Sampling Using JAGS

Finally, and exceptionally for this book, we shall be using the JAGS-software to do the Gibbs sampling for us. The downside of JAGS, and similar software, is that it does not show its actual sampling algorithm, so you cannot improve that where necessary. However, for statistical models that can be expressed in the JAGS language (which excludes, for example, process-based models), JAGS often works very well. The following JAGS-code, which can be run from R if the package `rjags` has been installed, is largely self-explanatory, but you can find more information in Chap. 17. It is also very fast and produces for our problem the results shown in Fig. 14.5.

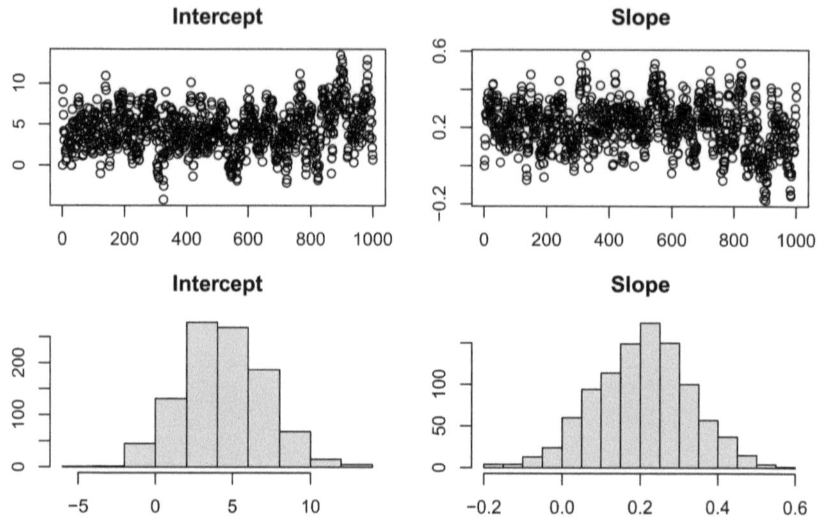

Fig. 14.4 Gibbs sampling using own R-code

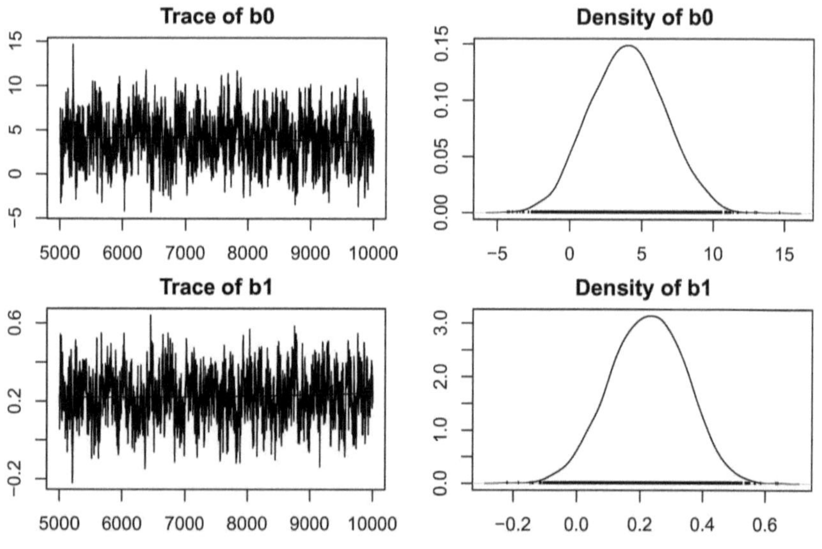

Fig. 14.5 Gibbs sampling using JAGS

```
f.JAGS <- " model { for (i in 1:n) {
    y[i]      ~ dnorm( y.hat[i], tau.y[i] )
    y.hat[i] <- b0 + b1 * x[i]
    tau.y[i] <- pow( sy[i], -1 ) }
  b0 ~ dnorm( 0, 1e-4 ) ; b1 ~ dnorm( 0, 1e-4 ) } "
writeLines( f.JAGS, con="f.JAGS.txt" )
f.JAGS.data <- list ( n=ny, y=y, x=x, sy=diag(Sy) )
f.JAGS       <- jags.model( "f.JAGS.txt", data=f.JAGS.data, n.adapt=5e3 )
update( f.JAGS, n.iter=5e3 )
f.JAGS.codaSamples <- coda.samples( f.JAGS, var=c("b0","b1"), n.iter=5e3 )
f.JAGS.mcmcChain   <- as.matrix( f.JAGS.codaSamples )
```

14.15 Comparison of Methods

So why do all the aforementioned methods give the same results? Well, the primary answer is that they are all correct ways of calculating the Bayesian posterior for the simple case of linear Gaussian modelling. Moreover, the eight analytical methods all effectively carry out regression using the same formula for a conditional Gaussian distribution. However, there are computational differences, in particular because of the different matrix inversions that the methods carry out.

The numerical, sampling-based methods are less accurate and precise for the given problem, but they have the major advantage of being applicable to far more problems, and we can always improve them by using longer chains, better proposal distributions or different MCMC algorithms. For the calibration of complex models, such as process-based models, we have no choice but to use numerical sampling algorithms, unless we are satisfied with coarse approximations to the posterior distribution.

Exercises

1. **Prior uncertainty.** Reduce the prior variances, which we specified as being 10^4 in Eq. (14.7), to more realistic values such as 10^2. Does that increase or decrease the posterior estimates for β? Why?
2. **LS72-equations.** We mentioned in the text that the LS72-equations are reduced to the GLS equations when you set the prior parameter uncertainty to very high values (i.e. set $\Sigma_\beta^{-1} = 0$). But what would the posterior uncertainty have been if you instead had set $\Sigma_y^{-1} = 0$)?
3. **Kalman Filtering (KF) without intercept uncertainty.** In our application of KF, we specified a very large prior uncertainty for β. Repeat the KF exactly as before, but with $\Sigma_\beta[1, 1]$ set to zero. How does that change the posterior distribution for β? What kind of linear regression have you just done? Why is the posterior uncertainty for the slope parameter lower than before?

Chapter 15
Gaussian Processes and Model Emulation

15.1 Model Emulation

Sampling-based estimation of the posterior distribution is computationally demanding. We have already mentioned the continuing search for efficient MCMC algorithms. MCMC is especially slow when the model of interest is a process-based model (PBM) with a long run-time. In such cases, it may be good to replace the PBM with a faster surrogate model. The surrogate model will take the same inputs as the original model but calculate the output more quickly. However, its output cannot be exactly the same as that of the original model, so it just provides an approximation. If the surrogate model is a statistical model that produces not just the approximative prediction of what the original model would have produced, but a whole probability distribution, then it is called a *statistical emulator*, or just *emulator* for short.

So an emulator is a replacement for the original model f. It is also a function of x and θ, but its output is not $f(x, \theta)$ itself but a probability distribution for it. So the emulator is a tool for predicting—probabilistically—the output from our model f without actually running the model. The idea is of course that we can build the emulator in such a way that it will be much faster to evaluate than the original model. The emulator $p_g[\,]$ is usually derived from a small training set of N modelling results $\{x_i, \theta_i, f(x_i, \theta_i)\}_{i=1..N}$:

$$p_g[\ g \mid x, \theta\ ] \ = \ p[\ f(x, \theta) = g \mid \{x_i, \theta_i, f(x_i, \theta_i)\}\ ]. \tag{15.1}$$

Because emulators are designed to compute faster than the original model, they are especially useful when we need a large sample of model runs. Bayesian calibration (using MCMC) is one example where we must evaluate a model many times to create a representative sample from the posterior distribution. However, emulator speed comes at the cost of only producing a probabilistic answer. This makes the likelihood function more complicated. Say our uncertainty about measurement error is captured by the parameter σ_y. Then the likelihood function for the original model

M. van Oijen, *Bayesian Compendium*,
https://doi.org/10.1007/978-3-031-66085-6_15

would be $L[\theta] = p[\epsilon_y = y - f(x, \theta)|\sigma_y]$, but when using the emulator, we would have to integrate out our uncertainty for g:

$$L'[\theta] = \int_{g=-\infty}^{\infty} p[\epsilon_y = y - g|\sigma_y]\, p_g[g|x, \theta]\, dg. \tag{15.2}$$

In Bayesian calibration, the posterior would then be proportional to $p[\theta]L'[\theta]$. To keep calculation of the likelihood-integral as well as $p[\theta]L'[\theta]$ manageable, emulators are generally selected for their mathematical properties. Gaussian process emulators are often used (Kennedy & O'Hagan, 2001). These approximate the original model using the multivariate normal distribution and use convenient functions for calculation of covariances.

15.2 Gaussian Processes (GP)

A *Gaussian Process (GP)* is a multivariate Gaussian distribution defined for infinitely many variates. Let's track back a bit here to explain what that means. The univariate Gaussian captures our uncertainty about a single quantity, and it is fully defined by just two parameters: its mean and variance. The bivariate Gaussian represents uncertainty about two quantities, so its definition requires us to quantify five parameters: two mean values, two variances and one covariance. A trivariate Gaussian has nine parameters: three mean values, three variances, three covariances. We see that the number of parameters for n-variate Gaussians increases as $n(n+3)/2$. So how can a GP, defined as a Gaussian where n is infinite, be of any practical use? Well, the solution is to define two functions: a *mean function* $m(s)$ and a *covariance function* $C(s, s')$ where the s and s' range over the infinite set of quantities. Both functions have their own parameters of course. Note that variances are covered by the covariance (or 'kernel') function because a variance is just the covariance of a quantity with itself ($C(s, s)$).

When do we use a GP? The typical case is when we are uncertain about a function $f(s)$, i.e. we do not know the values of $f(s)$ for most or all of the values of its arguments s. We can have that uncertainty because the function is nature itself, e.g. when we are interested how, in a region with spatial coordinates $s = (longitude, latitude)$, some soil property $f(s)$ varies. Our knowledge of soil properties and our understanding of soil properties will often be incomplete in which case we may decide to represent our uncertainty about soil property variation as a GP. Using a GP for such problems is called *Gaussian process regression* (GPR) or *kriging*, and it can also be considered a form of *machine learning* (see Chap. 21). GP can likewise be used if f is a model rather than nature itself. Say we are uncertain about the outcomes of a computer model f across parameter space $\{\theta \in \Theta\}$, and the model is too slow to run for many values of θ. Expressing our uncertainty about model outputs $f(s)$ using a GP is called model *emulation*, which is the subject of this chapter. Finally, the $f(s)$ can involve both nature and modelling. For example, we may use a GP to represent our uncertainty about how much the model outputs differ from reality for different model inputs (covariates x and/or parameters θ). In such a case, we are using a GP to represent the *discrepancy*. Other possible uses of GP are for representing variation in measurement error or model-data mismatch as a function of covariates: the possible uses of GP are endless.

Let's summarise. We denote a GP for a function of interest f as $GP[m(s), C(s, s')]$, often shortened as $GP[s]$, where $m(s) = E[f(s)]$ and $C(s, s') = \text{Cov}[f(s), f(s')]$. The GP is a multivariate normal distribution that is fully defined by is mean function $m(s)$ and its covariance function $C(s, s')$, defined for all s and s' in the domain S. The following two equations are then equivalent ways of saying that we use a GP emulator to represent our uncertainty about model output $f(s)$:

EMULATION USING A GAUSSIAN PROCESS:

$$f(s) \sim GP[m(s), C(s, s')] \tag{15.3}$$

$$f(s) = m(s) + T(s), \quad \text{where } T(s) \sim GP[0, C(s, s')].$$

For more details about GP and their connections with various statistical methods, see Kennedy and O'Hagan (2001) and Conti and O'Hagan (2010).

It is common to build emulators for *scalar* output from models, e.g. the final yield or average growth rate. So if the model that we want to emulate generates predictions for n different variables, then we could decide to build n different emulators, one for each of the outputs. And if each output variable comes as a predicted time series, then each individual emulator could represent, for example, the time-averaged output for a single variable (Andrianakis et al., 2015). For time series or spatially gridded output, an index of time or space could be added to the inputs of the emulator (Kennedy & O'Hagan, 2001). For another (Bayesian hierarchical) approach to combining scalar emulators to represent spatiotemporally distributed model output, see Leeds et al. (2013).

The more elegant—but difficult to implement—idea of directly emulating *multivariate* output from models is a matter of ongoing research (Conti & O'Hagan, 2010; Drignei, 2017; Higdon et al., 2008; O'Hagan, 2006; Rougier, 2008). Inspiration can be taken from recent developments in multivariate GP modelling of spatial processes that keep computational demand in check by representing inter-variable independencies in the cross-covariance functions (Dey et al., 2022). For other approaches than GP in emulation, see, e.g. Hauser et al. (2012).

A note of caution about the literature. Earlier in this book, we objected to the term *random variable* because nothing is really random in nature. There is only our own uncertainty, and what we represent by means of probability distributions is just that—our uncertainty—and not some intrinsic randomness. The terminological inconsistency is most stark when modellers define their model parameters to be *constants* yet when they try to estimate their values consider them to be random *variables*! Likewise, we should object to the use of the term *random function* to denote a function whose behaviour we are uncertain about. This is just a heads-up for when you start studying the literature on emulation, where an emulator is often described as being a random function.

15.3 Emulating a One-Input, One-Output Model

We shall now be building a GP emulator for a very simple model, which has only one input variate and one output variate. So our model is a function f that takes scalar input x and provides one matching scalar output $y = f(x)$. Say that we have run the model for three values of x, and we have collected the inputs and outputs in vectors $x = \{x_i\}$ and $y = \{y_i\}$ for $i = 1..3$. Let's now use these training data to build a GP emulator of the model and use the emulator to estimate the function for values of x that were not evaluated.

```
x <- c(10,20,30) ; y <- c(6.09,8.81,10.66)
```

We start with the simple case where we have values for all GP parameters except the regression parameters β of the mean function. In other words, we have the covariance function of our GP but not its mean function. For the covariance function in this example, we choose an overall variance $V_y = 3$, and we assume that the covariance between points decreases as a negative exponential function of distance with a correlation length $\phi = 10$. So we find the covariance matrix for the training data with the following code:

```
Vy <- 3 ; phi <- 1e1          # GP variance and correlation length
dx <- as.matrix( dist(x) ) # Distance matrix
Sy <- exp( -dx/phi ) * Vy   # Covariance matrix
```

That only leaves the prior for the regression parameters to be specified. We have two regression parameters (an intercept and a slope), so the design matrix X for the regression will consist of two columns: one filled with all 1's, the other with x. For the parameter pair, we specify a prior mean vector μ_β and a prior covariance matrix Σ_β, which we denote in R-code as mb and Sb.

```
mb <- c( 0, 0 ) ; Sb <- diag( 1e4, 2 )
```

Just to remind you of how the design matrix X works: the matrix–vector product $X\,\mu_\beta$ is a vector consisting of the prior regression values for all model input values x. At this point, we can choose two ways to continue: we can implement the GP for this model ourselves using known analytical solutions, or we can use an R-package to build the GP for us. We show both ways but will check that they give the same results. The R-package (we choose geoR) will turn out to be useful both for producing plots and for trying out other covariance functions than our main choice, the negative exponential. But let's begin with showing the analytical solutions explicitly.

15.3.1 Analytical Formulas

When, as in this example, the covariance matrix Σ_y has been assigned (Sy in the R-code), the posterior distribution for the regression parameters can be found analytically (Lindley & Smith, 1972) as we already showed in Chap. 14:

GP with only regression parameters unknown:

$$
\begin{aligned}
& p[\beta|y] = N[\mu_{\beta|y}, \Sigma_{\beta|y}], \text{ where} \\
& \Sigma_{\beta|y} = (\Sigma_\beta^{-1} + \mathbf{X}^\top \Sigma_y^{-1} \mathbf{X})^{-1}, \\
& \mu_{\beta|y} = \Sigma_{\beta|y}(\Sigma_\beta^{-1}\mu_\beta + \mathbf{X}^\top \Sigma_y^{-1} y).
\end{aligned}
\tag{15.4}
$$

These equations can be programmed as follows:

```
GP.est <- function( x, y, Sy, mb, Sb, X=cbind(1,x) ) {
  Sb_y <- solve( solve(Sb) + t(X) %*% solve(Sy) %*% X )
  mb_y <- Sb_y %*% ( solve(Sb) %*% mb + t(X) %*% solve(Sy) %*% y )
  return( list( "mb_y"=mb_y, "Sb_y"=Sb_y ) ) }
```

And with our choices for data and prior this gives the following results for the posterior mean and covariance matrix of the regression parameters of this GP:

$$
\mu_{\beta|y} = \begin{bmatrix} 3.9068 \\ 0.2286 \end{bmatrix}; \quad \Sigma_{\beta|y} = \begin{bmatrix} 6.742 & -0.259 \\ -0.259 & 0.013 \end{bmatrix}. \tag{15.5}
$$

At this point, we have fully defined and trained our GP. Our mean function is $X\beta$ with parameter uncertainty $N[\mu_{\beta|y}, \Sigma_{\beta|y}]$ and our covariance function is the negative exponential $C[s, s'] = V_y \exp(-d(s, s')/\phi)$ with $V_y = 3$ and $\phi = 10$. So we can now study our model function f, which is unknown apart from the training data, by means of this emulator instead of f itself.

But how exactly do we use the emulator to predict the output from f for a point x_0 where we do not have the function evaluation $f(x_0)$? Well, because the GP is a well-behaved multivariate normal distribution, the predictive distribution $N[\mu_{y_0|y}, \Sigma_{y_0|y}]$ can be found analytically too. Essentially it boils down to: (1) finding the distances from x_0 to all training locations $\{x_i\}$, (2) using the distances to calculate the covariances $\{C(x_0, x_i)\}$, and (3) using the covariances to determine how much $\mu_{y_0|y}$ and $\Sigma_{y_0|y}$ differ from, respectively, $X_0\beta$ and V_y. The R-code for the analytical solution is as follows:

```
GP.pred <- function(x0,x,y,Sy,phi,mb_y,Sb_y,X0=c(1,x0),X=cbind(1,x)) {
  dx0  <- if( is.vector(x) ) { abs(x-x0)
          } else { sapply(1:length(y),function(i){dist(rbind(x0,x[i,]))}) }
  C0   <- Sy[1] * exp( -dx0/phi )
  m0_y <- X0 %*% mb_y - t(C0) %*% solve(Sy) %*% (X %*% mb_y - y)
     a <- X0 - t(C0) %*% solve(Sy) %*% X
  S0_y <- Sy[1] - t(C0) %*% solve(Sy) %*% C0 + a %*% Sb_y %*% t(a)
  return( list( "m0_y"=m0_y, "S0_y"=S0_y ) ) }
```

(The ugly `if – else` construction that we use in this code to calculate the vector of distances to the target point `x0` allows the function also to be used when there is more than one coordinate, i.e. when x is not a vector but a higher-dimensional array. This will be useful later.)

Before we start using these predictive equations, note that for both the posterior mean function and the posterior covariance function, we need to invert the prior covariance matrix Σ_y (Sy in the R-code). So when we choose a covariance function for our GP, we do not have complete freedom: we must choose a function that leads to a proper invertible symmetric covariance matrix. In mathematical parlance, we must ensure that the matrix is 'positive semidefinite' (Rasmussen & Williams, 2006). But that still leaves much choice, such as exponential, Gaussian (squared exponential), spherical and other covariance functions. We shall explore a few of these functions later in this chapter.

Let's now calculate results of our emulator for the as yet unevaluated points $x_0^a = 10.1$ and $x_0^b = 15$. The first is very close to a point in our dataset ($x[1] = 10$), the second is not close. Our R-function GP.pred gives the following results for predictive mean and variance at these two points:

```
> x0=10.1: mean = 6.117023 ; variance = 0.05926102
> x0=15.0: mean = 7.437081 ; variance = 1.410466
```

As expected, the predictive mean for the first point is very similar to the nearby observation value ($y[1] = 6.09$), and it has low uncertainty, whereas the second point has a mean predictive value that is roughly the average of its nearest neighbours, but those are not nearby so the predictive variance is very high.

Notice that in the example that we have just given, the x played a double role, as coordinates and as covariate values. We used the x as coordinates to calculate distances between points, and from those distances we calculated the covariances. And we used the x as covariate values in our regression. We shall often see GP with this double role of x, especially in emulation. But it is quite possible to have a GP where the x are coordinates and nothing more, e.g. when we just want to estimate a global mean without any regression on covariates. In that case the covariate matrix X would just consist of a single column with the value 1 in each position, and there would be no columns with coordinate values. And on the other hand, it is also possible to have a GP where for each location we have more covariates than coordinates, e.g. in some forms of geostatistical kriging.

15.3.2 *R-package* geoR

The R-package 'geoR' requires specifying two spatial coordinates, typically longitude and latitude. But by specifying the same latitude for all points, we mimic a one-input model, which is what we need here for our simple univariate model. We now use the package to find the posterior for the intercept and slope and then calculate the marginal predictive distributions for the same two points as earlier (one close to a measured point and one further away). Here is the R-code:

```
s.11     <- as.geodata( cbind( x, rep(0,3), y) )
xpred    <- c(10.1,15) ; xpred.11 <- expand.grid( xpred, 0 )
model.11 <- model.control(cov.m  ="exponential",
                          trend.d=~coords[,1], trend.l=~ xpred.11[,1])
prior.11 <- prior.control(beta.prior     ="normal", beta     =mb,
                          beta.var.std=Sb/Vy,
                          sigmasq.prior  ="fixed" , sigmasq  =Vy,
                          phi.prior      ="fixed" , phi      =phi,
                          tausq.rel.prior="fixed" , tausq.rel=0)
out.11   <- output.control(messages=FALSE)
geoR.11  <- krige.bayes(s.11,loc=xpred.11,mod=model.11,pr=prior.11,out=out.11)
```

The posterior parameter distribution found by geoR is the following:

```
> Mean: 3.906838 0.228601
> Variance:
>            beta0      beta1
> beta0  6.7424941
> beta1 -0.2592242 0.01296323
```

The parameter estimates from geoR are, as we would hope, the same as we found with the direct implementation of the theoretical equations. And now let's examine geoR's prediction for $x_0^a = 10.1$ and $x_0^b = 15$:

```
> Means: 6.117023 7.437081 , Variances: 0.05926102 1.410466
```

The prediction results are also the same as we found analytically. So we can rely on geoR whenever that is more convenient than implementing the theory ourselves. Let's now conclude this section by using geoR to calculate the predictive distribution for many more points. We can then plot these to show the shape of the emulated function across its domain, with uncertainty bounds. We will first do so using the exponential covariance function that we have used all the time up to now, but we shall also show the consequences for the GP interpolation of using ten different covariance functions. The results are shown in Fig. 15.1.

We see that the choice of covariance function generally matters little for the posterior mean function (the unbroken lines in the figure) but has major impact on

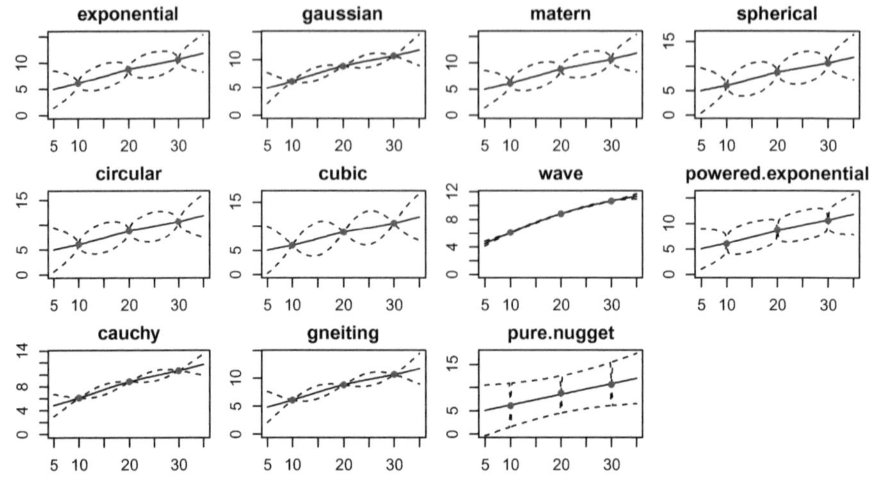

Fig. 15.1 GP emulation of a 1D model y=f(x) by means of geoR: results for 11 different covariance functions

our predictive uncertainty away from training points. Note also that the last panel differs from the others. It shows the results from geoR's 'pure.nugget' covariance function, which actually represents no spatial covariance at all, just local noise. If you choose any of the other covariance functions but let the spatial correlation length ϕ approach zero, your results will be the same as for 'pure.nugget'.

15.4 Emulating a PBM (EXPOL6)

The remainder of this chapter will be concerned with GP emulation of model EXPOL6. This is a 6-parameter version of the expolinear model (Goudriaan & Monteith, 1990) that we introduced and calibrated in Chaps. 8 and 11. The model produces output of biomass W and leaf-area index LAI as a function of time t and 6 model parameters. We want to make an emulator for the LAI-predictions. For the emulation, we shall only use our own equations as geoR cannot go beyond two dimensions as far as the parameter covariance matrix is concerned.

We shall be working with the same prior distribution for the model that we specified in Chap. 11. To make the GP calculations computationally robust, we use a 'scaled' version of the model, which we call EXPOL6s, where the magnitude of every parameter is around 1. Within the R-definition of this modified model, the dimensionless scaled parameter values are transformed back to their original units. Note from this R-code that also the covariate t (which represents time and is the first input to the model) is scaled, in its case by division by 100.

```
EXPOL6s <- function( t=0, b=c(1,1,1,1,1,1) ) {
  t      <- 100     * t
  I0     <-  10     * b[1] # MJ PAR m-2 ground d-1
  K      <-   1     * b[2] # m2 ground m-2 leaf area
  LAIMAX <-   3     * b[3] # m2 leaf area m-2 ground
  LAR    <-   0.007 * b[4] # m2 leaf area g-1 DM
  LUE    <-   2     * b[5] # g DM MJ-1 PAR
  W0     <-   1     * b[6] # g DM m-2 ground
  e1     <- exp(I0*K*LAR*LUE*t) ; e2 = exp(K*LAR*W0) ; e3 = exp(-K*LAIMAX)
  W      <- W0*e3 + ( log(e2+1/e1-1)/(K*LAR) + I0*LUE*t ) * (1-e3)
  LAI    <- log( (1+e1*(e2-1)) / (1+e1*(e2-1)*e3) ) / K
  return( list( "W"=W, "LAI"=LAI ) ) }
```

15.4.1 Training Set

We need a training set of completed EXPOL6s calculations to calibrate our GP-emulator on. So we sample from the prior parameter distribution for EXPOL6s by means of *Latin Hypercube sampling*, which ensures good coverage of prior parameter space (O'Hagan, 2006). And for each parameter vector, we store the model output at multiple times. We make all combinations of parameter vectors (nl = 9) and times (nx = 5), giving us a total of n = 45 different input vectors with corresponding EXPOL6s outputs.

```
set.seed(13)
nl        <- 9 ; lhs6 <- lhs::randomLHS( n=nl, k=nb )
sample.b  <- t(sapply( 1:nl, function(i){bmin6 + lhs6[i,]*(bmax6-bmin6)} ))
sample.x  <- c(20,40,60,80,100) / 100 ; nx <- length(sample.x)
sample.xb <- NULL
for (x in sample.x) {sample.xb <- rbind( sample.xb, cbind(x,sample.b) ) }
```

We now run the model for each of the 45 input vectors to create the output vectors (W, LAI), which are shown in Fig. 15.2. This completes our training dataset.

15.4.2 Calibration of the Emulator

We assume values for the hyperparameters of the covariance function: Vy = 3 and phi = 0.2. And we choose a very wide prior for our eight regression parameters which are: the intercept, the slope for the single covariate (time) and the six slopes for the regular parameters of EXPOL6s. Then we calibrate the (regression part of the) GP as we did above for the other examples. That gives us the following values for the posterior means and standard deviations of the eight parameters.

```
> mb_y = ( -7.376 3.013 0.6362 1.784 1.22 1.924 1.936 -0.4282 )
> Standard deviations:  55.05 1.046 52.49 1.872 0.9928 3.09 1.271 1.312
```

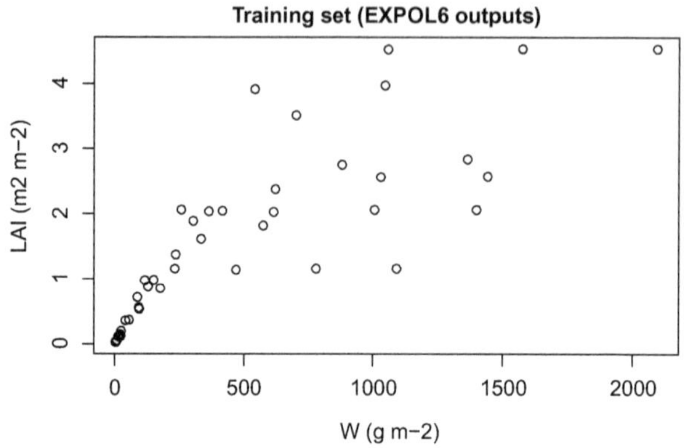

Fig. 15.2 Training set for GP-emulation of EXPOL6: Outputs from 45 input vectors

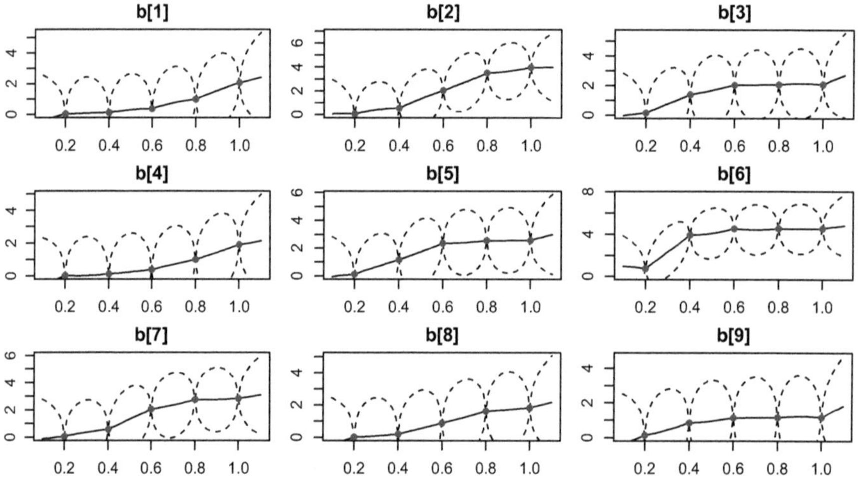

Fig. 15.3 In-sample emulation of EXPOL6: interpolation of LAI ($m^2\ m^{-2}$) vs. scaled time for the 9 parameter vectors that were used in GP calibration. Blue dots are the training points, i.e. the LAI-output values from EXPOL6

15.4.3 *Testing the Emulator*

Now we want to see how our emulator performs for prediction. We begin with checking the in-sample prediction, where we look at the emulator's interpolation capability for the parameter vectors that were present in the training set of inputs. Remember that the training was only for five times, so what we check here is how the emulator performs for those same parameter vectors at other times. The results of the in-sample prediction are shown in Fig. 15.3.

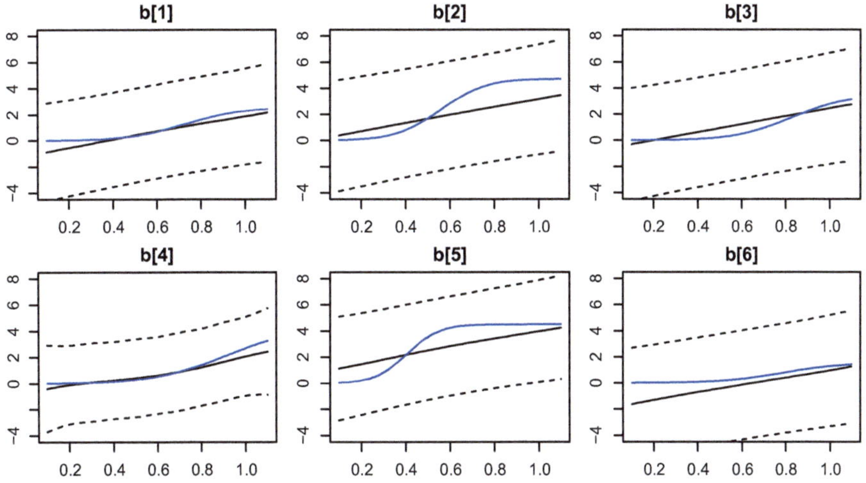

Fig. 15.4 Out-of-sample emulation of EXPOL6: interpolation of LAI ($m^2\ m^{-2}$) vs. scaled time for 6 parameter vectors that were not used in GP calibration. Blue line: EXPOL6. Black line and dashed lines: emulator

The lines connect the dots, so the emulator gives the correct results for the training inputs, but we see that uncertainty at intermediate times quickly becomes large. Now let's check the emulator for six parameter vectors randomly drawn from the prior distribution and not present in the training data. We expect that the emulator's mean function will not correspond to the true values from EXPOL6. The results are shown in Fig. 15.4. We see that the emulator is always close to the actual EXPOL6 curves, but the emulator does not reproduce the pronounced S-shaped LAI-growth curves that EXPOL6 produces with the second and fifth parameter vector.

We test emulator output more thoroughly for t = 0.1, 0.4, 0.7 and 1.0. For each of those times, we compare model and emulator output for 1000 different parameter vectors. Note that usually we shall not be able to run such extensive tests: emulators are made when the original model is very slow to run. The results are shown in Fig. 15.5. We see that the emulator differs most from the original model for t=0.1 and performs poorly for t=1.0. These are times where the EXPOL6 model behaves most non-linearly: the time series produced by EXPOL6 for the LAI are exponential in the beginning, then become linear and end asymptotically. Note, for example, that EXPOL6 can never produce negative outputs, but the emulator does, especially for the earliest times.

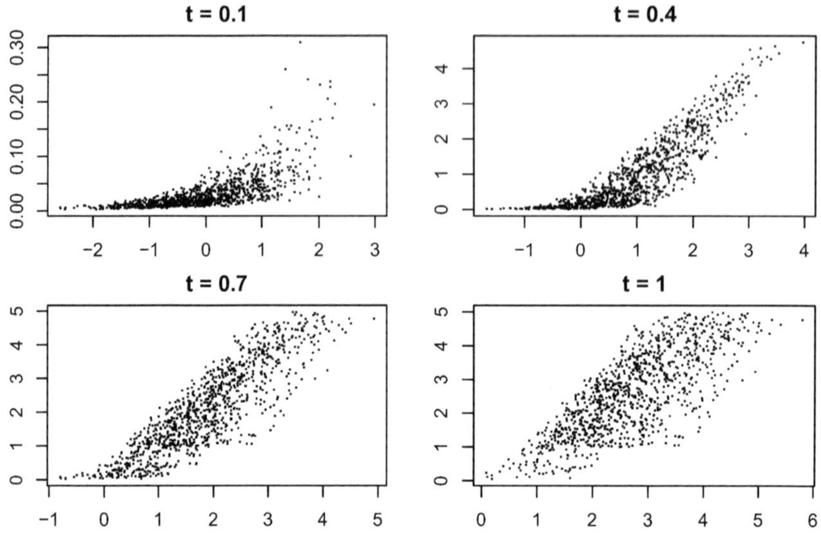

Fig. 15.5 Out-of-sample emulation of EXPOL6: estimation of LAI (m^2 m^{-2}) at 4 scaled times for 1000 parameter vectors that were not used in GP calibration. x-axis: emulator, y-axis: EXPOL6

15.5 Improving the Emulation

Our two examples of model emulation in this chapter have been simplistic in that we assumed that the covariance function hyperparameters (such as the correlation length constant ϕ in the exponential covariance function) were known. Obviously, these hyperparameters should be estimated rather than guessed, and we discuss the impact of hyperparameter uncertainty in the chapter on spatial modelling (Chap. 23). `geoR` can estimate hyperparameters, but the package cannot be used for emulating functions with more than two parameters. More flexible R-packages for estimating GP-parameters are `BACCO` and `DiceKriging` (Roustant et al., 2012).

Besides that, you will probably have many more ideas on how to improve our emulators. We made assumptions regarding the shape of the mean- and covariance functions that could be changed. It pays off to spend considerable time in building your emulator and all the techniques that statisticians have developed for improving model fit are options: improve the design of the training set, develop the emulator piece-wise for different parts of its domain, transform the output from the original model (e.g. a log-transformation for the early phase or a logistic transformation for the whole function domain), etc. Tokmakian et al. (2012) discuss methods for emulation of complex nonlinear models, and fortunately their examples suggest that we can be optimistic about the performance of carefully built GP-based emulators provided enough training data can be generated by the to-be-emulated model. Andrianakis et al. (2015) and Loeppky et al. (2009) discuss how the required number of model runs used for building an emulator should increase with the number of model parameters.

Our emulators were built for deterministic models, so they reproduced model outputs for training data exactly, as Fig. 15.3 showed. However, it can be advantageous to add a so-called *nugget* term—a random noise term $\epsilon \sim N[0, \sigma_\epsilon^2]$—to the emulator, which was found by Andrianakis and Challenor (2012) to improve the numerical stability of the emulation. The term 'nugget' comes from geostatistical techniques originally developed for the mining industry.

The GP approach to emulator building is empirical and not mechanistic. Hybrid approaches that combine statistical (including machine learning) and mechanistic approaches to emulator building have been studied by Carbajal et al. (2017). It may also be useful to explore ideas from the field of *systems identification* which aims for dynamic surrogate models. An example is the work by Young (1998) who used transfer functions to derive low-dimensional surrogate models that replaced complex climate simulators with systems of just three differential equations which—in stark contrast to GP-emulators—remained interpretable in terms of real-world processes.

Exercise

1. **GP with small correlation length.** Use the `geoR`-code of this chapter to show that a GP with a linear trend-term and a very small correlation length (ϕ) is equivalent to linear regression as far as estimation of the regression parameters is concerned.

Chapter 16
Graphical Modelling

A *graphical model* (GM), also called a *probabilistic network*, is a representation of a joint probability distribution. A GM has two parts: (1) a graph with nodes connected by edges and (2) information about the nodes. So the graph is just the visible part of the model. GMs do not represent a new kind of statistical model; they are just helpful tools for analysing joint probability distributions. Every distribution can be represented by a GM, so whatever your research problem or modelling method is, you can choose to use a GM to organise your thinking. And you can choose from multiple kinds of GM as we show in Fig. 16.1. The two most prominent types are *Bayesian networks* (BN, which use directed graphs so that the edges are arrows) and *Markov Random Fields* (MRF, which use undirected graphs) (Briganti et al., 2023). Both classes can represent continuous as well as discrete probability distributions. The different types of GM all come with their own apparatus for designing, interpreting and updating the distributions that they represent. The graphs play an important role in that they indicate ways of decomposing the distribution into conditionally independent parts. Bayesian calibration of a GM consists of acquiring data for one or more nodes (the others remain unobserved *latent variables*) and using Bayes' Theorem to update the joint probability distribution. We leave this very short general introduction to GM here to concentrate on *Gaussian Bayesian Networks* (GBNs), which can showcase most of the advantages that GM bring. But we shall finish this chapter with some general comments on GM.

M. van Oijen, *Bayesian Compendium*,
https://doi.org/10.1007/978-3-031-66085-6_16

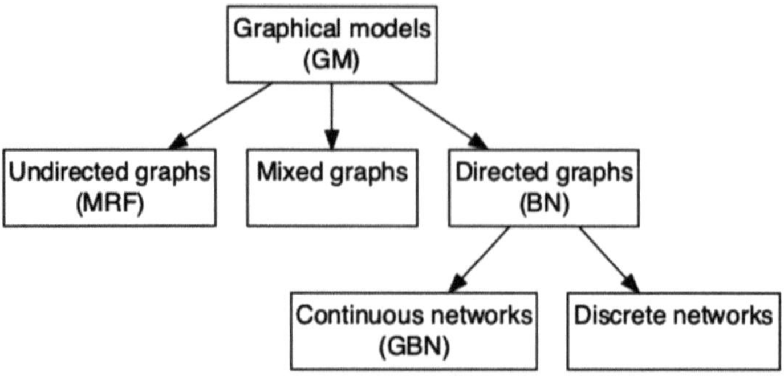

Fig. 16.1 Simplified scheme of graphical modelling categories with only the major types shown

16.1 Gaussian Bayesian Networks (GBN) and Causality

A *GBN* is a multivariate Gaussian probability distribution, but one for which we have taken the trouble of providing a nice depiction in the form of a *Directed Acyclic Graph (DAG)*. Every multivariate Gaussian is fully specified by its mean vector and covariance matrix, but these are not shown in the graph. Instead, the graph shows how the distribution can most efficiently be factorised. GBNs are a subset of *Bayesian Networks* (BN, also called *Bayesian Belief Networks*, BBN). BNs can represent every joint probability distribution, including ones for which some or all of the marginal distributions are not continuous as in a GBN, but discrete.

The DAG for the simplest GBN has two nodes linked by one edge (Fig. 16.2). The nodes represent the variables we are uncertain about, and the edge shows how we propose to decompose their joint distribution.

For such a simple GBN, factorisation of the joint probability distribution, $p[A, B]$, is trivial. The arrow in the figure goes from A to B, which indicates that $p[A, B]$ is to be factorised as $p[A]\, p[B|A]$, but the opposite factorisation, $p[B]\, p[A|B]$, is mathematically equivalent and may be just as simple. It is only when our networks are more complicated that some factorisations may be simpler (have fewer edges) than others. For such networks, the ordering of nodes we show in the graph, using the arrows, becomes important. If we let the arrows represent causality, i.e. running from causes to effects, we generally produce the simplest DAG.

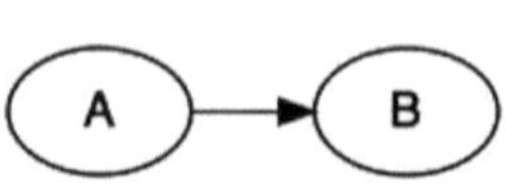

Fig. 16.2 The simplest Directed Acyclic Graph (DAG)

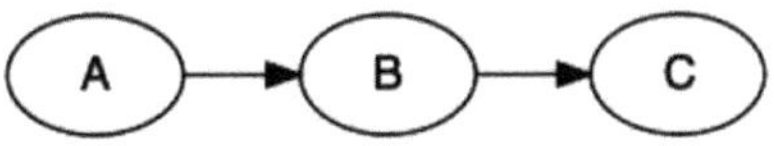

Fig. 16.3 C is conditionally independent of A

16.1.1 Conditional Independence

Say we have a network with three nodes A, B and C. Then we can always write the joint pdf as $p[A, B, C] = p[A]\, p[B|A]\, p[C|A, B]$, or the same equation with any other permutation of A, B and C. The corresponding graph would have three arrows: from A to both B and C, and from B to C. But what if we knew that the relationship between A and C was fully mediated by B? For example, A = resources, B = growth rate, C = yield, where we assume that any correlation between resources and yield is fully due to resources stimulating growth and that in turn leading to high yield. In such a case, we say that C is conditionally independent of A. Our DAG would then have just two arrows (Fig. 16.3).

This DAG represents our belief that $p[A, B, C]$ can be factorised simply as $p[A]\, p[B|A]\, p[C|B]$. The factorisation implied by the graph may be the most natural (it is a chain of cause and effect), but other factorisations are possible that represent exactly the same joint probability distribution. This three-node probability chain is the simplest example of a *hierarchical model* (the subject of the next chapter).

16.2 Three Ways to Specify a Multivariate Gaussian

GBNs represent multivariate Gaussians graphically, but how do we specify the underlying probability distribution quantitatively? The most common way is to give the mean vector and the covariance matrix (or its inverse, the *precision matrix*). The diagonal of the covariance matrix gives the *unconditional variances* of the nodes, which quantify our uncertainty about the node values when the values of the other nodes are not known yet. Our knowledge about each node is then not yet constrained, or 'conditioned', by knowledge of the other nodes.

But sometimes a quite different specification is more useful where we say how much uncertainty we still have about each node after the values of all other nodes have in fact become known. That is done by giving the vector of *conditional variances*. That vector of course tells us nothing about the relationships between the nodes, but if we complement it by also giving the mean vector and the regression coefficients for each pair of nodes, then we have again fully determined the distribution underlying our GBN.

To summarise, if we have a GBN with n nodes and r edges, there are three different ways of specifying its multivariate Gaussian distribution:

1. $n \times 1$ mean vector + $n \times n$ covariance matrix,
2. $n \times 1$ mean vector + $n \times n$ precision matrix,

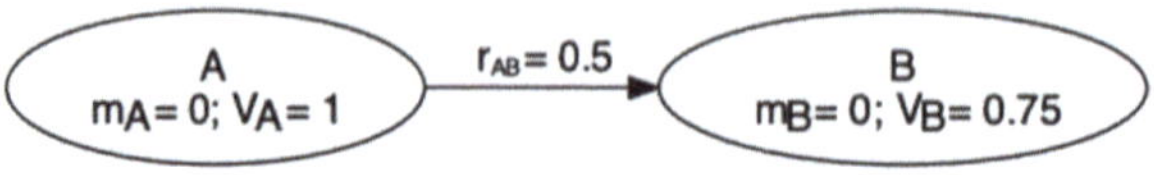

Fig. 16.4 DAG with means and conditional variances specified in the node ellipses and edge labelled with the regression coefficient

3. $n \times 1$ mean vector + $n \times 1$ vector of conditional variances + $r \times 1$ vector of regression coefficients.

All three methods have their uses and will appear in this chapter. But note that only the third method makes use of the information in the DAG, and it is also often the easiest to specify as we shall see. And the vector information that it provides could be shown in the DAG itself, which would make the DAG fully self-explanatory. All we need to do for that is adding the mean and conditional variance to each node and labelling each edge with the regression coefficient. See Fig. 16.4 for an example.

A convenient property of GBNs is that any two nodes for which the regression coefficient is zero are conditionally independent. That tells us that we do not need to draw an arrow between those nodes. In fact, the main use of the graph is to show which nodes are conditionally independent: absence of an arrow tells us more than the presence of one.

16.2.1 Switching Between the Three Specifications

It is easy to switch between the three ways of specifying the multivariate Gaussian. Conversion between (1) and (2) is conceptually the easiest: the covariance and precision matrices are each other's inverse. So, if the two matrices are denoted as Σ resp. W, then $W = \Sigma^{-1}$, which in R-code would be written as `solve(Sigma)`.

The conversion from (3) to (2) can be done using the algorithm of Shachter and Kenley (1989), which we code in R as follows:

```
precMatrix <- function( nodes, Vcond, R ){
  n <- length(nodes) ; W <- 1 / Vcond[1]
  for(k in 2:n){
    rk       <- R[ k, 1:(k-1) ]
    W_top    <- cbind( W * Vcond[k] + rk %*% t(rk), -rk ) / Vcond[k]
    W_bottom <- cbind(                          -t(rk),   1 ) / Vcond[k]
    W        <- rbind( W_top, W_bottom ) }
  rownames(W) <- colnames(W) <- nodes
  return(W) }
```

This function has three arguments: `nodes`, `Vcond` and `R`. The first argument is the vector of node names. The second is an equally long vector of conditional variances.

And the third is a matrix R which has zeroes everywhere except for the elements R_{ij} that correspond to an edge from node j to node i in the DAG. Let's apply this to the example DAG of Fig. 16.4. The following three lines of code give the three different specifications. The code does not specify the mean vector, but it is the same for all three methods, and we choose $\mu = (0, 0)$.

```
nodes <- c("A","B") ; Vcond <- c(1,0.75) ; R <- matrix( c(0,0.5,0,0), nrow=2)
W     <- precMatrix( nodes, Vcond, R )
Sigma <- solve(W)
```

When we run this code, we find that the distribution implied by the DAG shown in Fig. 16.4 has the following mean vector, covariance matrix and precision matrix:

$$\mu = \begin{bmatrix} 0 \\ 0 \end{bmatrix}; \quad \Sigma = \begin{bmatrix} 1 & 0.5 \\ 0.5 & 1 \end{bmatrix}; \quad W = \begin{bmatrix} 1.333 & -0.667 \\ -0.667 & 1.333 \end{bmatrix}. \tag{16.1}$$

This shows that the conditional variance of B that was specified in the DAG ($V_B = 0.75$) is less than its unconditional variance in the covariance matrix ($\Sigma[2, 2] = 1$) because knowledge of A reduces uncertainty about B, thanks to the non-zero regression coefficient ($R[2, 1] = 0.5$). In contrast, the conditional variance of A in the DAG ($V_A = 1$) is the same as its unconditional variance ($\Sigma[1, 1] = 1$) because we use the factorisation of the joint probability distribution where A is a parent of B and not the other way around. Note that this lack of symmetry between A and B is in no way inherent or visible in our mean vector and covariance matrix. It is simply the result of us specifying an ordering by means of the DAG, which defines how the regression coefficients are calculated—in this case for B as a function of A and not vice versa.

Castillo et al. (2008) provided an algorithm for the conversion from method (2) to (3) that we implement here as follows:

```
VcondR <- function( W ) {
  n  <- dim(W)[1] ; Vcond <- rep( NA, n )
  Wk <- W         ; R     <- matrix( 0, nrow=n, ncol=n )
  for(k in n:2){
    ik       <- 1 : (k-1)
    Vcond[k] <- 1 / Wk[ k, k]
    R[k,ik]  <-    -Wk[ k,ik] * Vcond[k]
    Wk       <-     Wk[ik,ik] - as.matrix(R[k,ik]) %*% R[k,ik] / Vcond[k] }
  Vcond[1] <- 1 / Wk[1,1]
  return( list( Vcond=Vcond, R=zapsmall(R) ) ) }
```

Earlier we derived the precision matrix from the DAG of Fig. 16.4. Let's verify that our new function gives us back the conditional variances and regression coefficients when we apply it to that precision matrix. So we run the command:

```
VcondR(W)
```

and that indeed retrieves the conditional variances and regression coefficients that we started out from:

$$\texttt{Vcond} = \begin{bmatrix} 1 \\ 0.75 \end{bmatrix}; \quad \texttt{R} = \begin{bmatrix} 0 & 0 \\ 0.5 & 0 \end{bmatrix}. \tag{16.2}$$

We now have the tools available for our study of GBNs in the remainder of this chapter.

16.3 The Simplest DAG Is the Causal One!

A good DAG that follows causal reasoning (with edges going from causes to consequences) will have fewer edges than non-causal DAGs for the same joint probability distribution. Let's study this with an example. We specify a causal DAG for the impacts of rain and irrigation on soil water based on the following code chunk:

```
nodes <- c("Rain","Irrigation","SoilWater")
m     <- c( 2    , 0.5          , 1        )
Vcond <- c( 0.2  , 0.05         , 0.1      )
R     <- matrix( rep(0,9), nrow=3) ; R[3,1] <- 0.9 ; R[3,2] <- 1
```

The causal DAG is drawn on the left in Fig. 16.5. It has three nodes but only two edges (non-zero values in the R matrix). If we wanted to make life difficult for

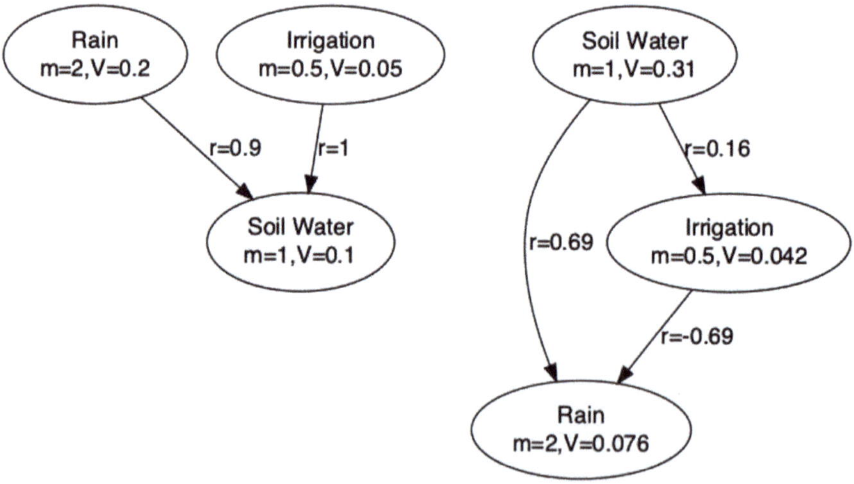

Fig. 16.5 Two DAGS for the same probability distribution. The causal DAG on the left is smaller than the non-causal one on the right

ourselves, we could: (1) Derive the covariance matrix for this DAG, (2) Reorder the mean vector to ('SoilWater', 'Irrigation', 'Rain') and reorder the covariance matrix accordingly, (3) Derive the conditional variances and regression coefficients for this new ordering. The non-causal DAG that follows from that exercise is shown in the same figure on the right.

The non-causal DAG has three edges, which is the maximum that a three-node DAG can have. So it is more complicated than the two-edges causal DAG, even though both DAGs represent the same joint probability distribution! Causal thinking helps us keep our DAGs simple, which makes them much easier to work with.

16.4 Sampling from a GBN and Bayesian Updating

Because a GBN is a probability distribution, we can sample from it. This can be done one node at a time, following the factorisation implied by the structure of the DAG, but we can also do this using standard algorithms for sampling vectors from the whole multivariate Gaussian in one go. The first method—which only involves conditionally *univariate* Gaussians—will be computationally more efficient when the network has many nodes because it avoids the inversion of the covariance matrix.

16.4.1 Updating a GBN with New Information

After we have specified our network, new information may become available. For example, the value of one or more nodes may become known through measurement or observation. When such new information becomes available, in any BBN, not just GBNs, then that information propagates to other nodes. If the *evidential nodes* have no *parents*, just *children*, then updating the network is just a matter of basic forward propagation of information following the direction of the arrows to all the *descendants* of the evidential nodes. The situation is more complicated if the evidential nodes themselves are children of other nodes. But then Bayes' Theorem can be used to update the distribution.

In the case of GBNs, Bayesian updating of the network is very easy: there is an analytical solution to the problem of finding the 'posterior network', so we need not worry about how to implement Bayes' Theorem. That analytical solution is implemented in R here. The algorithm operates on the covariance matrix of the GBN so for this task we need to provide the covariance specification of the GBN. (There are computationally more efficient ways that take advantage of the network structure, but those take more steps and are not needed for the simple examples of this chapter.) If all we have is the specification in terms of the conditional variances and regression coefficients, then we must first convert that specification into the covariance matrix (via the precision matrix) with the code shown above. Here is code for the posterior mean and covariance matrix of a multivariate Gaussian distribution after we learn the values y for one or more of the nodes.

```
GaussCond <- function( mz, Sz, y ) {
  i <- 1 : ( length(mz) - length(y) )
  m  <- mz[i]   + Sz[i,-i] %*% solve(Sz[-i,-i]) %*% (y-mz[-i])
  S  <- Sz[i,i] - Sz[i,-i] %*% solve(Sz[-i,-i]) %*% Sz[-i,i]
  return( list( m=m, S=S ) ) }
```

In this implementation, the vector y (with length n_y) represents the nodes that become known from, for example, measurement. The code assumes that the measurements correspond to (and replace) the last n_y values of the mean vector. The function returns the new mean vector and covariance matrix for the smaller network that remains when the known nodes are removed and the others updated. Let's use it for a very simple example, using the same 2-node DAG as earlier (Fig. 16.4), where we now assume that we learn that B=2. The function will thus return the posterior mean and variance of A. You can code this as `GaussCond(c(0, 0), Sigma, 2)`, which will tell you that $p[A|B] = N[1, 0.75]$. Compare that posterior conditional mean and variance for A with the numbers in the DAG of Fig. 16.4. Are the results as you expected?

16.5 Example I: DAG Design, Sampling and Updating

We shall now illustrate some uses of graphical models with the example of a 4-node GBN. We start from the mean + covariance matrix specification:

```
nodes <- c( "A", "B", "C", "D" )
m     <- c(  3 ,  4 ,  9 ,  14 )
S     <- matrix( c(4 ,  4,  8, 12,
                   4 ,  5,  8, 13,
                   8 ,  8, 20, 28,
                   12, 13, 28, 42), nrow=4 )
```

We now calculate the conditional variances and the regression coefficients to prepare us for plotting the DAG:

```
VR <- VcondR( solve(S) ) ; Vcond <- signif(VR$Vcond,2) ; R <- VR$R
```

When we inspect R, we see that there are only four non-zero regression coefficients. So we need only four edges in our DAG. It would have been difficult to conclude directly from the covariance matrix that there are so many conditional independencies in this GBN! The DAG is shown in Fig. 16.6.

If we want to assess the uncertainty represented by this 4-node GBN, we should sample from it. And because we know the mean vector `m` and covariance matrix `S`,

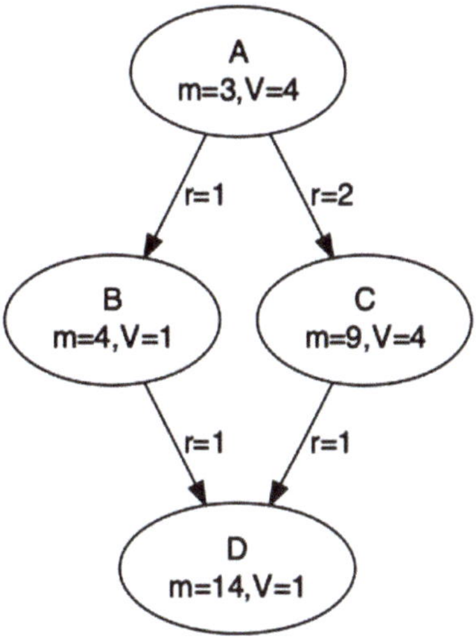

Fig. 16.6 DAG with four nodes

we could sample with the R-command `rmvnorm(100, m, S)`. Because our GBN does not have many nodes, that will not be computationally demanding. However, let us instead apply the method that is more suitable for large GBNs, namely sequential sampling, one node at a time. To do that efficiently, we follow the ordering of the nodes in the DAG and make use of the regression coefficients, as in the following code:

```
n <- 100 ; sampleSeq <- sapply( LETTERS[1:4], function(x) NULL )
sampleSeq$A <- m[1] + rnorm( n, 0, sqrt(Vcond[1]) )
sampleSeq$B <- m[2] + rnorm( n, 0, sqrt(Vcond[2]) ) +
                      R[2,1] * (sampleSeq$A - m[1])
sampleSeq$C <- m[3] + rnorm( n, 0, sqrt(Vcond[3]) ) +
                      R[3,1] * (sampleSeq$A - m[1])
sampleSeq$D <- m[4] + rnorm( n, 0, sqrt(Vcond[4]) ) +
                      R[4,2] * (sampleSeq$B - m[2]) +
                      R[4,3] * (sampleSeq$C - m[3])
lapply( sampleSeq, mean )
```

You can verify for yourself that this produces a representative sample from the multivariate Gaussian of the GBN.

Now let's assume that the first node, A, becomes known exactly through measurement. Say A = 7. What does that mean for the remaining nodes? Well, we defined the function `GaussCond <- function(mz, Sz, y)` that gives us the posterior multivariate Gaussian when one or more nodes becomes known. Unfortunately, the function definition requires the known nodes to be the *last* ones in the specification, and our node A is the first. So we need to reorder the mean vector and covariance matrix by putting A at the end before applying the function.

```
i2 <- c(2:4,1) ; nodes2 <- nodes[i2] ; m2 <- m[i2] ; S2 <- S[i2,i2]
GaussCond(m2,S2,7)
> $m
>      [,1]
> [1,]    8
> [2,]   17
> [3,]   26
>
> $S
>      [,1] [,2] [,3]
> [1,]    1    0    1
> [2,]    0    4    4
> [3,]    1    4    6
```

It is also possible to provide measurements from multiple nodes at the same time. The following code chunk shows how to do this, but we leave running the code to the reader. The code shows the case where the measurements are {A=7, B=8, C=17}.

```
i3 <- c(4,1:3) ; nodes3 <- nodes[i3] ; m3 <- m[i3] ; S3 <- S[i3,i3]
GaussCond(m3,S3,c(7,8,17))
```

If the values for the evidential nodes would have been different, the posterior mean would have been different too, but the posterior variance would have been the same.

16.6 Example II: A Linear Chain

In this example, only neighbours in the chain $[A] \to [B] \to [C] \to [D] \to [E]$ are not conditionally independent.

```
n <- 5                         ; nodes <- LETTERS[1:n]
m <- rep(0,n)                  ; Vcond <- rep(1,n)
R <- matrix(0,nrow=n,ncol=n) ; for(j in 2:n){ R[j,j-1] <- 0.5}
```

The precision matrix of such a linear chain is characterised by having zero's everywhere except on the middle three diagonals:

```
precMatrix( nodes, Vcond, R )
>       A     B     C     D    E
> A  1.25 -0.50  0.00  0.00  0.0
> B -0.50  1.25 -0.50  0.00  0.0
> C  0.00 -0.50  1.25 -0.50  0.0
> D  0.00  0.00 -0.50  1.25 -0.5
> E  0.00  0.00  0.00 -0.50  1.0
```

16.7 Example III: Univariate Linear Dependency

The relationships in a GBN are linear: all children depend linearly on their parents. The slopes of the linear relationships are given by the regression coefficients R_{ij}. The strengths of the relationships can be assessed by inspecting the covariance matrix. Some nodes may be so closely correlated that the relationship is effectively a deterministic linear function as Example III will show. After that, Example IV will demonstrate that GBNs can also represent multivariate stochastic linear relationships.

To show that relationships in a GBN can be nearly deterministic, let's construct a 2-node GBN with a near-deterministic relationship, given as $y = 4 - 0.22x + \sigma_\epsilon$, with $\mu_x \sim N(8, 1)$ and $\sigma_\epsilon^2 = 10^{-6}$. We encode this as a bivariate Gaussian in the following way:

```
a  <- 4 ; b  <- -0.22
mx <- 8 ; Vx <- 1 ; my <- a + b * mx ; m <- c( mx, my ) ; Ve <- 1.e-6
S  <- matrix( c(  Vx, b*Vx, b*Vx, b^2*Vx + Ve), nrow=2 )
```

We sample from this multivariate Gaussian to show that we have indeed represented the intended near-deterministic dependency of y on x (Fig. 16.7).

Now let's represent this relationship as a GBN. For that we need the conditional variances of x and y and the regression coefficient. We calculate them in the usual way from the covariance matrix:

```
VR <- VcondR( solve(S) ) ; Vcond <- signif(VR$Vcond,2) ; R <- VR$R
```

We shall not plot the DAG, but it would simply consist of two nodes $[x] \to [y]$, with mean values (8, 2.24), conditional variances $(1, 10^{-6})$, and with regression coefficient -0.22. Note that in this simple case, we could have inferred the conditional variances and the regression coefficient directly from our premises $(y = 4 - 0.22x + \sigma_\epsilon, \sigma_x^2 = 1, \sigma_\epsilon^2 = 10^{-6})$. But the main point of this section is to show that GBNs, though limited in that all conditional probabilities must be Gaussian, can include (nearly) deterministic influences of some nodes on others.

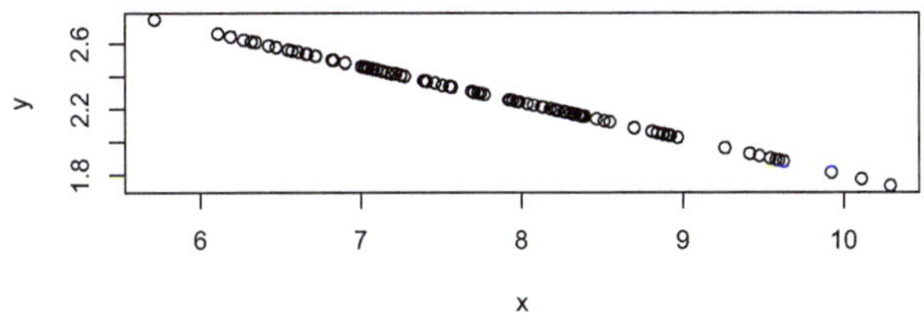

Fig. 16.7 Sampling from a multivariate Gaussian that mimics a near-deterministic linear relationship

16.8 Example IV: Multivariate Stochastic Linear Relations

This time we show that GBNs can also represent a linear sum of terms. We'll make the relationship more strongly stochastic this time, so not nearly deterministic. Say $z = 2 + 3x + 5y$, with $\mu_x = 10$, $\mu_y = 1$ and $V_x = V_y = 1$. We assume no linear relationship between x and y.

```
a  <-  2 ; b1 <- 3 ; b2 <- 5
mx <- 10 ; Vx <- 1 ; my <-  1 ; Vy <- 1 ; mz <- a + b1 * mx + b2 * my
m  <- c( mx, my, mz ) ; Ve <- 10
S  <- matrix( c(     Vx,       0, b1  *Vx,
                      0,     Vy,          b2*  Vy,
                 b1*Vx, b2*Vy, b1^2*Vx + b2^2*Vy + Ve ), nrow=3 )
```

So we now constructed the mean vector and covariance matrix of the multivariate Gaussian representing $z = 2 + 3x + 5y$ with the given variances. We sample from it and (for once) use R's 'lm'-function to confirm that we have produced data from the desired relationship.

```
n          <- 1e5
samplexyz <- rmvnorm( n, m, S )
samplex <- samplexyz[,1]
sampley <- samplexyz[,2]
samplez <- samplexyz[,3]
lm( samplez ~ samplex + sampley )
>
> Call:
> lm(formula = samplez ~ samplex + sampley)
>
> Coefficients:
> (Intercept)      samplex      sampley
>       2.148        2.987        4.990
```

16.9 Example V: Geostatistical Interpolation

In our final example of graphical modelling, the nodes now represent n different points in space. Our prior uncertainty about the node values will be represented by a multivariate Gaussian with covariances that depend on the distances between points. We shall use a negative exponential covariance function with a given correlation length ϕ, i.e. $\mathrm{Cov}(i, j) = \exp(-\mathrm{distance}(i, j)/\phi)$. We choose the mean vector to be equal to 2 everywhere, and the variances equal to 1.47. We assume that there are $n = 4$ points of interest, named 's1' to 's4' and located at coordinates (0,3), (4,0), (4,3) and (0,0) respectively. We assume that $\phi = 1.9115$. That defines our multivariate distribution as follows:

```
n    <- 4 ; nodes <- paste0( "s", 1:4 ) ; m <- rep( 2, n ) ; phi <- 1.9115
dist <- matrix( c( 0, 5, 4, 3,
                   5, 0, 3, 4,
                   4, 3, 0, 5,
                   3, 4, 5, 0 ), nrow=4 )
S    <- 1.47 * exp(-dist/phi)
```

In a typical geostatistics problem, we get measurement values for all nodes except one. The problem is then to infer the posterior distribution for that unmeasured node. Once we know how to do that, we can move our unmeasured location to any location in our region of interest, get our estimates for that point and so on. That way we can build up a map for the whole region when only a finite number of points have been measured. That method is called *Gaussian process regression* (GPR) or *kriging*. And in this chapter, we shall be using a graphical model for that. So let's make a GBN, with its DAG, for our spatial network of nodes (Fig. 16.8).

```
VR    <- VcondR( solve(S) )
Vcond <- signif(VR$Vcond,2) ; R <- signif(VR$R,2)
```

Now let's assume that we measure values at three of the locations: {s1=0.7, s2=3.1, s3=2.2}. So we want to estimate the probability distribution for s4. Let's calculate the posterior distribution for s4 in the same way as we have done before in this chapter.

```
  i2 <- c(4,1:3) ; nodes2 <- nodes[i2] ; m2 <- m[i2] ; S2 <- S[i2,i2]
  y  <- c(0.7,3.1,2.2) ; GaussCond(m2,S2,y)
> $m
>          [,1]
> [1,] 1.862632
>
> $S
>          [,1]
> [1,] 1.387985
```

Fig. 16.8 DAG for kriging

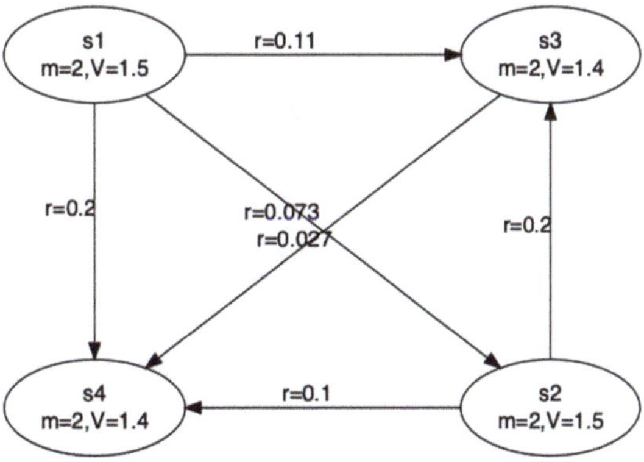

So the kriging prediction for point s4 was a mean value of 1.86 with a kriging variance of 1.39. This was of course a very simple example of geostatistics, where we assumed that a priori the mean value everywhere was 2, independent of the coordinates. More usually, in geostatistics, we first fit a trend surface to the data and then model the residuals using the multivariate Gaussian. For that approach, see Chap. 23.

16.10 Applications of Graphical Modelling

Graphical modelling is a widely applicable tool for formulating and visualising statistical models. All statistical models can be formulated as a joint probability distribution that we depict in a graph, but only some have a tradition of doing so. These include *structural equation modelling*, *path analysis*, *Hidden Markov Models* (HMM) and *hierarchical modelling* which is the subject of the next chapter. A graphical model that presents the general structure of *Bayesian Decision Theory* (BDT) will be shown in Chap. 19.

Our examples in this chapter were all GBNs, which made it easy to specify the conditional probability distributions for each node. But Bayesian networks are in fact more often used to depict quite irregular discrete distributions, where each node comes with its own 'Conditional Probability Table' (Hojsgaard et al., 2012). This is partly because commercial software for GM design and calibration can currently not handle continuous distributions. For a tutorial on such networks, with code examples in R, see Briganti et al. (2023).

Discrete Bayesian networks have frequently been used in environmental management and have shown their use for facilitating stakeholder involvement (Gonzalez-Redin et al., 2016; Smith et al., 2012). People can co-design the network by discussing the variables and influences that the graph needs to represent. The discussions focus on representing causality which keeps the graphs sparse and understandable. Contrast this with our first example earlier which showed that the causality-agnostic covariance matrix looked much more complicated than the graph we derived from it. The graph revealed many unsuspected conditional independencies.

Exercises

1. **Bayes' Theorem as a DAG.** If you wanted to show Bayes' Theorem as a two-node DAG, would you choose $[\theta] \rightarrow [y]$ or $[y] \rightarrow [\theta]$?
2. **Covariance matrices (after MacKay).** Which of the following four matrices could be the covariance matrix of a GBN with three edges: $[y1] \rightarrow [y2]$, $[y1] \rightarrow [y3]$, $[y2] \rightarrow [y3]$?

```
M1 = matrix( c(9,  3, 1,  3,  9,  3, 1,  3, 9), nrow=3 )
M2 = matrix( c(8, -3, 1, -3,  9, -3, 1, -3, 8), nrow=3 )
M3 = matrix( c(9,  3, 0,  3,  9,  3, 0,  3, 9), nrow=3 )
M4 = matrix( c(9, -3, 0, -3, 10, -3, 0, -3, 9), nrow=3 )
```

3. **Simplest hierarchical model.** Consider a population of people with three binary properties: age, smoking habits and whether they are runners or not. A person with $(\xi, \theta, y) = (1, 1, 1)$ is someone who is old, a smoker, and a runner. And $\xi = 0$ indicates them not being old, etc. We assign the simplest hierarchical model $[\xi] \rightarrow [\theta] \rightarrow [y]$ (compare Fig. 16.3) with the following conditional probability distributions: $p[y|\theta = 0] = \text{Br}[0.4]$, $p[y|\theta = 1] = \text{Br}[0.1]$, $p[\theta|\xi = 0] = \text{Br}[0.2]$, and $p[\theta|\xi = 1] = \text{Br}[0.8]$. (Remember that we use Br[0.4] as shorthand notation for a Bernoulli distribution with 40% probability of 1 and 60% probability of 0, and so on.) Our prior probability for ξ is $Br[0.5]$, so even odds for a person being old or young.

- **a.** What is the likelihood for an old person being a runner? In other words, what is $p[y = 1|\xi = 1]$?
- **b.** And what is it for a young person, i.e. what is $p[y = 1|\xi = 0]$?
- **c.** You are going to meet Jan, who is a runner. What is the posterior probability that Jan is old, i.e. what is $p[\xi = 1|y = 1]$? Use Bayes' Theorem to answer this.
- **d.** Generate 10^4 populations from this graphical model, each with 20 people. What percentage of the populations represents Simpson's Paradox, where $p[y = 1|\xi = 0] > p[y = 1|\xi = 1]$, while $p[y = 1|\xi = 0, \theta = 1] < p[y = 1|\xi = 1, \theta = 1]$ and $(p[y = 1|\xi = 0, \theta = 0] < p[y = 1|\xi = 1, \theta = 0]$? In other words, in how many populations is the percentage who runs highest amongst the young, while within both subgroups of smokers and non-smokers it is highest amongst the old?

Chapter 17
Bayesian Hierarchical Modelling

In the previous chapters, our statistical procedure was very simple: define a prior probability distribution for the parameters $p[\theta]$ and a likelihood function $L[\theta] = p[y|\theta]$, and that was it. Bayes' Theorem then told us what the posterior distribution would be once we received the data: $p[\theta|y] \propto p[\theta]L[\theta]$. The prior for the parameter vector was always a fully specified distribution, e.g. the product of known univariate Gaussians. In *Bayesian hierarchical modelling* (BHM), we do not specify the prior that directly. Instead we make the prior distribution depend on other parameters, which we call *hyperparameters*. Here is a table of the differences:

	Non-hierarchical	Hierarchical
Likelihood	$p[y\|\theta]$	$p[y\|\theta]$
Prior	$p[\theta]$	$p[\theta\|\xi]$
Hyperprior	–	$p[\xi]$

(17.1)

Strictly, hierarchical models are nothing new because they can always be collapsed into non-hierarchical models. You do that by combining the parameters and hyperparameters in a new longer parameter vector $\omega = (\theta, \xi)$ and finding the 'flattened', non-hierarchical prior distribution by multiplication: $p[\omega] = p[\theta|\xi]p[\xi]$.

When we carry out Bayesian calibration of the parameters and hyperparameters in a hierarchical model, statisticians refer to the approach as Bayesian hierarchical modelling (BHM, Cressie & Wikle, 2011). In BHM, we must specify a prior for the hyperparameters, $p[\xi]$, which then automatically implies a prior for the regular parameters, $p[\theta|\xi]$. But the likelihood function does not change in this hierarchical setup, because model output $f(x, \theta)$ only depends on the regular model parameters, not the hyperparameters. So, in BHM, the posterior distribution is defined as:

$$p[\theta, \xi|y] \propto p[\xi]\, p[\theta|\xi]\, p[y|\theta] = p[\xi]\, p[\theta|\xi]\, L[\theta]. \tag{17.2}$$

M. van Oijen, *Bayesian Compendium*,
https://doi.org/10.1007/978-3-031-66085-6_17

Note that the use of BHM implies greater data needs. For example, if the hyperparameters in the BHM stand for spatial variability, then calibration cannot be effective without data from dispersed sites.

17.1 Why Hierarchical Modelling?

So when would this decomposition of the prior be useful? Well, it turns out that it is often very convenient to think hierarchically. The parameters θ can represent properties of individuals, and the hyperparameters ξ properties of the groups to which the individuals belong (Cressie et al., 2009; Ogle, 2009; Ogle & Barber, 2008). For example, modellers need not assume that all trees in a region have the same wood density but can allow for spatial variation represented by two hyperparameters: mean wood density and its standard deviation. So they would represent the wood density of n individual trees as $\theta = (\theta_1, \ldots, \theta_n)$ and the mean and variance of those wood densities as $\xi = (\mu, \sigma^2)$.

In disciplines such as ecology, hierarchical modelling is rapidly becoming the standard approach. Ecosystem modellers are giving increased attention to how various plant and animal traits vary with environmental conditions (Simpson et al., 2016). Hierarchical modelling allows for great flexibility in such cases: each parameter can range from being completely generic (zero inter-group variance) to being completely site- or species-specific (large inter-group variance). A good introduction to hierarchical modelling in the context of vegetation research was provided by Dietze et al. (2008). They quantified variability in parameters for tree allometry (power law relationships for height vs. crown area) both within and between species in US forests and showed how even species for which few data were available could be fit reliably within the hierarchical model.

But the scope for using BHM goes well beyond representing spatial or taxonomic differences: the many different sources of error mentioned in this book can all be represented explicitly, with their own hyperparameters, in a hierarchical statistical model (Cressie et al., 2009). For example, BHM facilitates the use of different datasets in one Bayesian calibration whereby differences in measurement precision and accuracy are represented by dataset-specific and generic error-parameters (Ogle, 2009). Bayesian hierarchical modelling is therefore also the obvious tool for *meta-analysis*, where we want to extract robust conclusions from multiple observational studies that partly, but not completely, share similar measurement methods (DuMouchel & Normand, 2000; Ogle et al., 2014; Williams et al., 2018).

17.2 Examples of Non-hierarchical and Hierarchical Models

Let's make it all less abstract by comparing three statistical models for linear regression: two non-hierarchical ones and a hierarchical model. The three models are depicted as graphical models in Fig. 17.1.

The data will be growth measurements on plants from different grassland cultivars (varieties of grass). So we must decide whether we want to lump all plants together or to recognise the cultivars as different groups. Our starting point is as usual Eq. (2.5): $p[y|\theta] = p[f(x,\theta) - y = \epsilon_y]$, where we now choose $f(x,\theta) = a + bx$ with $\theta = (a,b)$. We expect data y from J different cultivars, with n_j measurements per cultivar, so $y = \{y_{ij}\}$ with $(i = 1..n_j,\ \ j = 1..J)$. We assume that measurement error has a Gaussian distribution $\epsilon_y \sim N[0, \sigma_y^2]$ with known variance σ_y^2. All this is quite standard. Our three models (A, B, C) only differ in their assumptions about the paramaters a and b. In model A, both parameters are generic, so they have the same value for all cultivars. In models B and C, the intercepts and slopes vary between cultivars. But only in model C are we going to estimate hyperparameters that represent the intercept- and slope variation between cultivars. So only model C will be a BHM. Despite these differences, the three models have, in a general sense, the same likelihood function:

$$p[y|\theta] = \prod_{j=1}^{J} \prod_{i=1}^{n_j} p[y_{ij}|\theta], \tag{17.3}$$

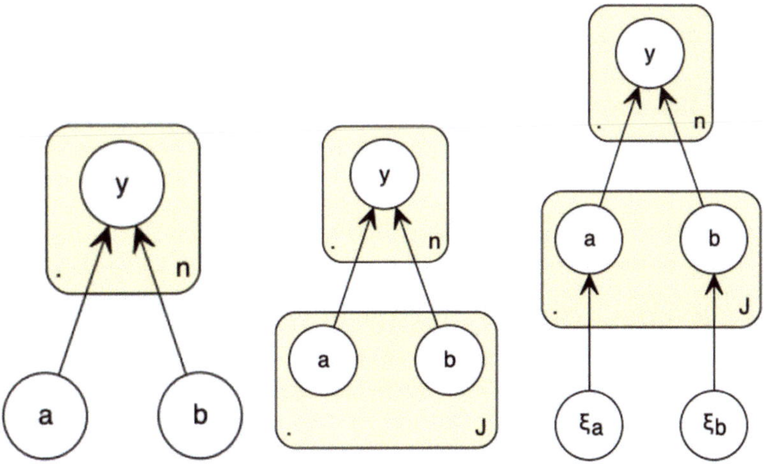

Fig. 17.1 Three statistical models for n observations on J different grassland cultivars. From left to right: Model A is a non-hierarchical model with universal slope ('a') and intercept ('b'); Model B is also non-hierarchical but has cultivar-specific parameters a and b; Model C is a Bayesian hierarchical model (BHM) with cultivar-specific parameters and global hyperparameters. We use plate-notation where the 'n' and 'J' in bottom-right corners indicate the multiplicity of data and cultivars

but the likelihood functions for individual data points $p[y_{ij}|\theta]$ do differ:

Model	Parameter vector	$p[y_{ij}\|\theta]$
A	$\theta = (a, b)$	$N[y_{ij}\|a + bx_{ij}, \sigma_y^2]$
B and C	$\theta = (a_1..a_J, b_1..b_J)$	$N[y_{ij}\|a_j + b_j x_{ij}, \sigma_y^2]$.

(17.4)

The priors for parameters (and hyperparameters in the case of model C) also differ. For the last model, we use four hyperparameters to represent the distributions of intercept and slope among the cultivars: $\xi = (\xi_a, \xi_b) = (\mu_a, \sigma_a^2, \mu_b, \sigma_b^2)$.

Model	Parameter vector	Prior
A	$\theta = (a, b)$	$p[\theta] = p[a]p[b]$
B	$\theta = (a_1..a_J, b_1..b_J)$	$p[\theta] = p[a_1]..p[a_J]p[b_1]..p[b_J]$
C	$(\theta, \xi) = (a_1..a_J, b_1..b_J, \mu_a, \sigma_a^2, \mu_b, \sigma_b^2)$	$p[\theta, \xi] = p[\theta\|\xi]p[\xi] =$
		$p[a_1\|\mu_a, \sigma_a^2]$ .. $p[a_J\|\mu_a, \sigma_a^2] \times$
		$p[b_1\|\mu_b, \sigma_b^2]$ .. $p[b_J\|\mu_b, \sigma_b^2] \times$
		$p[\xi_a]\ p[\xi_b]$.

(17.5)

Let's choose Gaussian priors for parameters and the two 'mean' hyperparameters, and uniform priors for the two 'variance' hyperparameters (not Gaussians because variances cannot be negative):

Model	Prior	
A	$p[\theta] =$	$N[a\|500, 10^6]\ N[b\|0, 10^2]$
B	$p[\theta] =$	$\prod_{j=1}^{J} N[a_j\|500, 10^6] N[b_j\|0, 10^2]$
C	$p[\theta, \xi] =$	$N[\mu_a\|500, 10^6]\ N[\mu_b\|1, 10^6]\ U[\sigma_a^2\|0, 10^3]\ U[\sigma_b^2\|0, 2] \times$
		$\prod_{j=1}^{J} N[a_j\|\mu_a, \sigma_a^2]\ N[b_j\|\mu_b, \sigma_b^2]$.

(17.6)

Models B and C are the most realistic models here, as they allow all parameters (a,b) to vary between cultivars. But if we get data, and the posterior distributions for σ_a^2 and σ_b^2 are such that only very small variances are plausible, then model A would be a good approximation. But we must be careful here. It may be tempting to begin by using the data y to find maximum likelihood estimates for the variances, i.e. $\widehat{\sigma_a^2}$ and $\widehat{\sigma_b^2}$, and then plug those estimates into the prior distributions for model A, i.e. $p[\theta] = p[a]\ p[b] = N[0, \widehat{\sigma_a^2}]\ N[0, \widehat{\sigma_b^2}]$. But if we then calibrate model A against the data using those priors, then we would be using the data twice, and that is illegal. The practice is called *empirical Bayes* and is of course not a proper Bayesian method, and it should be avoided.

Let's reflect some more about the differences between the three models. In model A, the parameters are generic, i.e. they are assumed to have the same value for all cultivars. If that is acceptable, then that would of course be very good, because our model could then be used everywhere without cultivar-specific recalibration. Model B is more modest, it assumes that intercepts and slopes are cultivar-specific. But it makes no assumptions about the variation of those parameters between cultivars. And what we learn about the intercept and slope of one cultivar, does not tell us anything about the value of these parameters for any other cultivar. Model C is much more useful in that respect, because if you calibrate that model, you are also learning about inter-cultivar variation in the parameter values. We can use that information about the hyperparameters (μ_a, σ_a^2) and (μ_b, σ_b^2) when we want to estimate slopes and intercepts for new cultivars. In short, hyperparameters define the prior for what to expect for the regular parameters (a, b) of new cultivars.

Let's now implement the three models. We shall be using real data on the growth of six different grassland cultivars. Our dataset contains 258 data points (t, y), where t is time in days, and y is plant dry matter (g m^{-2}). The number of data points for the cultivars $\{n_j\}$, $j = 1, .., 6$ is $\{10, 42, 80, 76, 36, 14\}$. The data are shown in Fig. 17.2.

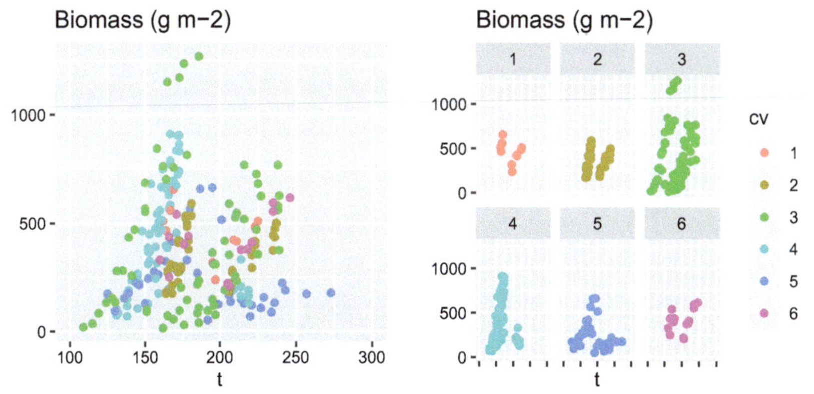

Fig. 17.2 Grassland growth data used for Bayesian calibration of three models. 't' = time (days), 'cv' = cultivar

17.2.1 Model A: Global Parameterisation

To define our models, we use the R-package `rjags`, which allows us to formulate the models in the 'JAGS' language for carrying out Bayesian calibration using MCMC. This goes against the promise that we would use basic R code where possible, but JAGS is convenient (especially for model C below) and you have seen the calibration of simple linear models with our own MCMC-code already in other chapters.

In JAGS, the first model is defined as follows:

```
ModelA <- " model {
  for (i in 1:ny) {
    y[i] ~ dnorm( a + b * t[i], tau.y[i] )
    tau.y[i] <- pow( sigma.y[i], -2 ) }
  a ~ dnorm( 5e2, 1e-6 ) ; b ~ dnorm( 1, 1e-2 ) } "
```

The main thing to notice is that the intercept *a* and slope *b* are not indexed, so they are universal parameters. The last line of the code shows the Gaussian priors for the two parameters. Note that JAGS has its own 'dnorm' command which requires specification of the *precision* rather than the *variance* as its second argument, so the small values we chose for the precision imply large prior uncertainty about both parameters.

To actually run the JAGS-code in our R-program, we need a few more lines of code. The main choices there are for the number of MCMC-chains to run (we choose 3) and for the lengths of the MCMC before and after burn-in (we choose 10,000).

```
writeLines( ModelA, con="ModelA.txt" )
data.A <- list ( ny=ny, y=y, t=t )
ModelA <- jags.model( "ModelA.txt", data=data.A, n.chains=3, n.adapt=1e4 )
update( ModelA, n.iter=1e4 )
ModelA.coda <- coda.samples( ModelA, var=c("a","b"), n.iter=1e4 )
```

17.2.2 Model B: Group-Specific Parameterisation

The code for model B is almost the same except that the parameters *a* and *b* are now indexed such that the model estimates their values separately for the $\texttt{ncv} = 6$ different cultivars.

```
ModelB <- " model {
  for (i in 1:ny) {
    y[i]      ~ dnorm( a[cv[i]] + b[cv[i]] * t[i], tau.y[i] )
    tau.y[i] <- pow( sigma.y[i], -2 ) }
  for (c in 1:ncv) { a[c] ~ dnorm( 5e2, 1e-6 ) ; b[c] ~ dnorm( 1, 1e-2 ) } } "
```

17.2.3 Model C: Hierarchical Parameterisation

Finally, we have the code for model C which begins the same as for model B but adds lines for the hyperparameters. Note that we use normal distributions for the hyperpriors of μ_a and μ_b and uniform distributions for σ_a^2 and σ_b^2.

```
ModelC <- " model {
  for (i in 1:ny) {
    y[i]      ~ dnorm( a[cv[i]] + b[cv[i]] * t[i], tau.y[i] )
    tau.y[i] <- pow( sigma.y[i], -2 ) }
  for (c in 1:ncv) { a[c] ~ dnorm( mu.a, tau.a ) ; b[c] ~ dnorm( mu.b, tau.b ) }
  mu.a ~ dnorm(5e2,1e-6) ; tau.a <- pow(sigma.a,-2) ; sigma.a ~ dunif(0,1e3)
  mu.b ~ dnorm(1,1e-6)   ; tau.b <- pow(sigma.b,-2) ; sigma.b ~ dunif(0,2) } "
```

17.2.4 Comparing Models A, B and C

We now look at the results from all three models. Figure 17.3 summarises the marginal posterior distributions for the parameters and (for model C) hyperparameters.

First of all, we notice that the posterior uncertainty for the two parameters of model A is very small, compared to the large variation in intercepts and slopes that we see for models B and C. This is because all 258 data points provide independent information for the estimation of just this single pair of parameters. We are reminded

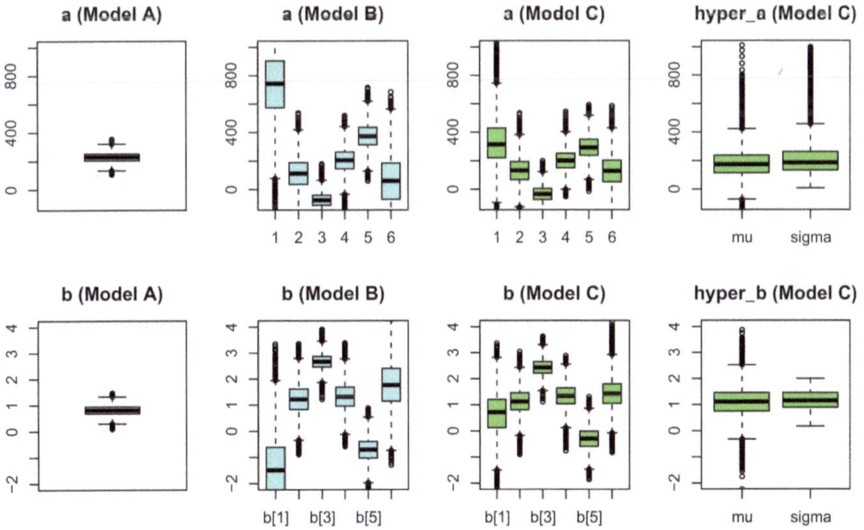

Fig. 17.3 Posterior parameter distribution for the three models

here that very low uncertainty about the parameters of a model does not mean that we can be confident that model structure is correct. Parameter uncertainty is generally smallest in oversimplified models. If we now compare the differences between cultivars, as determined by models B and C, we see much greater variation (especially for the intercept parameter a) in the non-hierarchical model. This is because model B is in fact a suite of six completely independent models, each of which only considers its own subset of data points. In the hierarchical model C, we see smaller variation between the cultivars: there is so-called *shrinkage* to the mean as defined by the hyperparameters ξ_a and ξ_b. This is typical behaviour of hierarchical models, and it is what we want: hierarchical modelling has in-built protection against outliers. The degree of shrinkage is highest for cultivars 1 and 6 because there were fewest observations for them (10 and 14, whereas all others had more than 35). So the BHM embodies the common sense that unusually high or low observations with poor replication are likely over- or underestimates. The estimates for cultivars 1 and 6 in the BHM also show lower posterior uncertainty, they have *borrowed strength* from the more frequently observed other cultivars.

17.3 Applicability of BHM

BHM, as represented by model C, is an extremely useful approach to modelling systems where the individual components or members are not completely unrelated and dissimilar (model B) but are also not perfectly identical (model A)—and that actually covers almost any system you may be working on! Bayesian hierarchical modelling should be the default approach to statistical modelling. It frees the modeller from the need to define as constants any parameters that in reality are variable, without running the risk of overparameterisation. For further reading, the books by Gelman et al. (2013) and Gelman and Hill (2006) are recommended. When delving into the literature, do not be discouraged by the usual Tower of Babel of confusing terminology. Hierarchical modelling is the same as *multilevel modelling* and *mixed modelling*, and our main example model C is a *linear mixed model* (LMM, see Chap. 20) that is also referred to as a *varying-intercept, varying-slope model.*

Exercises

1. **Taking the H out of BHM.** In Chap. 14, we showed 13 different ways of carrying out a linear regression. Could we have added BHM to the list? What kind of prior distribution for hyperparameters would make a BHM essentially identical to a non-hierarchical model?
2. **Simplest hierarchical model.** For some practice with hierarchical modelling of a discrete joint probability distribution, see the last exercise of the chapter on graphical modelling (Chap. 16).

Chapter 18
Probabilistic Risk Analysis

As Bayesians, we try to acknowledge all our uncertainties about data and models and express them as probability distributions. As we have seen, this approach allows us to quantify predictive uncertainty when we use our models for prediction. And this is of course important for the user of these predictions, whether that user is us or someone to whom we report our results. Our probabilistic predictions allow calculation of risks and, more generally, provide support for decision-making. This chapter introduces a rigorous method of probabilistic risk analysis (PRA), and the next chapter broadens the perspective to Bayesian decision theory (BDT). The two chapters summarise new theory for PRA and its links to BDT that was presented in greater detail by Van Oijen & Brewer (2022).

18.1 Risk, Hazard and Vulnerability

Risk is commonly defined as the expectation value for loss, irrespective of the system of interest. This means that we always distinguish two conditions: one where the system is not affected by some *hazard* and one where the system is being damaged and incurs a loss. Most risk theory is developed for discrete hazards that cause sudden system failure. Theoretical advances for systems where the hazard-determining variable is always present and continuously varying, with matching continuous system response, are less well developed (Van Oijen & Zavala, 2019). Take the example of dynamic systems like crops or cities whose properties change over time. The performance of such systems varies with resource availability and other environmental constraints. For example, crop growth depends on water supply, and urban health is a function of air pollutant concentration. Risks posed by

M. van Oijen, *Bayesian Compendium*,
https://doi.org/10.1007/978-3-031-66085-6_18

such environmental variables are not associated with sudden discrete events, but with extended periods of hazardous conditions (typically when the environmental variable has exceeded a threshold). To manage the risks, we need to know whether we should aim to reduce the *probability of hazardous conditions* or the *vulnerability of the system*.

Let's introduce notation. We denote risk as R, the probability of hazardous conditions as $p[H]$, and the vulnerability of the system as V. We need these notions to be defined in such a way that we can decompose the risk, i.e. write $R = p[H]\,V$. That decomposition clarifies how risk depends on both the probability of hazardous conditions and the damage then done. When both are high, so is the risk. Our definitions will be in terms of a system variable z and an environmental variable x. The definition of R is straightforward, as we follow the convention of equating it with expected loss: $R = E[z|\neg H] - E[z]$. Roughly, risk is optimal system performance minus average system performance. The definition of $p[H]$ is also obvious: $p[H] = p[x \in H]$. So it is the probability of environmental conditions being in the hazardous domain. That only leaves the definition of V which, to allow risk decomposition, we must define as $V = R/p[H]$. Expanding R according to its definition and simplifying give us $V = E[z|\neg H] - E[z|H]$. (We leave the details of the derivation as an exercise for the reader.) So vulnerability is the difference between optimal and suboptimal system performance. In summary, these are our basic equations for risk analysis:

PRA: Basic Equations

$$\begin{aligned} p[H] &= p[x \in H], \\ V &= E[z|\neg H] - E[z|H], \\ R &= E[z|\neg H] - E[z] \\ &= p[H]V. \end{aligned} \tag{18.1}$$

We have shown that vulnerability V is uniquely defined as the difference between two conditional system expectation values, for non-hazardous and hazardous conditions. And that definition follows directly from the standard definitions for R and $p[H]$. Yet, surprisingly, there is a general confusion in the risk literature about what 'vulnerability' means (Schneiderbauer & Ehrlich, 2004). Most of the different definitions of vulnerability that have been proposed do not allow formal risk decomposition. V is instead given a subjective scale such as 1 to 3, where '1 = low', '2 = medium' and '3 = high', and then placed against some measure of $p[H]$ in so-called 'risk matrices'. Risk matrices are used often, for example, in aviation and government (HM_Government, 2020). However, the approach is coarse, only semi-quantitative and does not rigorously scale risk as the expectation of loss (Thomas et al., 2014). We strongly recommend the use of our more formal probabilistic risk analysis which allows decomposition of risk into a term associated with the environment ($p[H]$) and one associated with the system (V).

18.2 Single-Threshold PRA

In risk analysis, the system variable z is typically univariate, such as growth or survival. The environmental variable x can be univariate (e.g. water availability or pollution level) or multivariate (e.g. rainfall plus wind speed, or rainfall at multiple times and locations). We shall keep things simple in this chapter and initially assume that hazardous conditions are those in which a univariate x falls below a threshold which we denote as thr. Imagine this as a resource that becomes limiting. So $p[H] = p(x < thr)$ and $p[\neg H] = p(x \geq thr)$ and so on. In this formalism, $p[H]$ is dimensionless, whereas V and R are physical variables that have the same dimension, e.g. kg m^{-2} s^{-1} or casualties per year. But the theory of this chapter will be applicable to all risk analysis when appropriate variable transformations are made, and we shall show some of that generalisation later.

With the assumption of a single hazardous region below a given threshold, our PRA-equations become:

Single-Threshold PRA

$$
\begin{aligned}
p[H] &= p[x < thr], \\
V &= E[z|x \geq thr] - E[z|x < thr]. \\
R &= E[z|x \geq thr] - E[z] \\
&= p[H]V.
\end{aligned}
\tag{18.2}
$$

In the paper that first introduced PRA as we develop it here (Van Oijen et al., 2013a), z stood for the productivity of vegetation, x was annual rainfall, and the hazard was drought defined as rainfall being less than its long-term 25th percentile. The probability of hazardous conditions (drought) was denoted as $p[H] = p[x < Q25]$ and the probability of no drought as $p[\neg H] = 1 - p[H] = p[x \geq Q25]$. We analysed risks from summer droughts to the productivity of vegetation across Europe under current and future climatic conditions (Van Oijen et al., 2014). This analysis showed that climate change will lead to greatest drought risks in southern Europe, primarily because of increased hazard probability but also because of changes in vulnerability (Fig. 18.1).

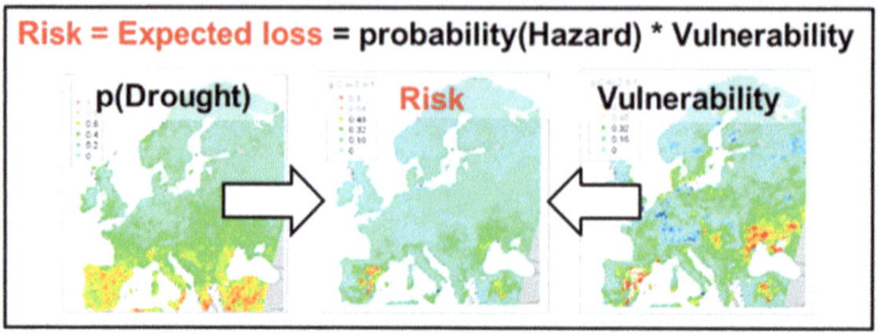

Fig. 18.1 Probabilistic risk analysis for drought impacts on European forest. Risks are highest in the Mediterranean area primarily because of greater hazard probability, but also because of greater vulnerability to drought. After Van Oijen et al. (2014)

18.3 Multiple-Threshold PRA

In *multiple-threshold PRA*, we distinguish $n \geq 2$ thresholds, with thr_1 being the smallest and thr_n the largest. This divides the hazardous region into n different intervals, and we carry out a separate PRA for each interval:

Multiple-Threshold PRA

$$\begin{aligned} p[H_i] &= p[thr_{i-1} \leq x < thr_i], \\ V_i &= E[z|x \geq thr_n] - E[z|thr_{i-1} \leq x < thr_i], \\ R_i &= p[H_i]\, V_i, \end{aligned} \tag{18.3}$$

where $i = 1, .., n$ and $thr_0 = -\infty$. Note that these equations for multiple-threshold PRA simplify to Eq. (18.2) when $n = 1$. But when $n \geq 2$ we can still retrieve Eq. (18.2) by summing the results for the different intervals as follows:

$$\begin{aligned} p[H] &= \sum p[H_i], \\ V &= \sum \frac{p[H_i]}{p[H]} V_i, \\ R &= \sum R_i, \\ &= p[H]\, V, \end{aligned} \tag{18.4}$$

with all summations running from $i = 1$ to n. So summarising multiple-threshold PRA in this way reduces the results to single-threshold PRA. The extra value of multiple-threshold PRA is that it gives us additional information about the relative probabilities and vulnerabilities for the n different sub-levels of the hazardous region.

18.4 Continuous PRA

In multiple-threshold PRA, we can specify as many discrete hazard levels as we want. In the limit of $n \to \infty$, we arrive at continuous PRA which only keeps the single highest threshold-level ($thr = thr_n$):

Continuous PRA

$$v(x) = E[z|x \geq thr] - E[z|x], \tag{18.5}$$
$$r(x) = p[x]\, v(x),$$

where $v(x)$, and $r(x)$ are continuous functions of the environmental variable x, just like the probability density function $p[x]$. And again we can retrieve the single-threshold PRA if we wish:

$$\begin{aligned} p[H] &= \int_{x=-\infty}^{thr} p(x)dx, \\ V &= \int_{x=-\infty}^{thr} \frac{p[x]}{p[H]} v(x)dx, \\ R &= \int_{x=-\infty}^{thr} r(x)dx, \\ &= p[H]\, V. \end{aligned} \tag{18.6}$$

18.5 Three Ways to Estimate the Expectations for z

We can now choose to use Eqs. (18.2), (18.3) or (18.5) to carry out our risk analysis. But how do we quantify the (conditional) expectations for z? We can do that in three different ways, depending on our prior knowledge. If all that we have is a sample of values (x_i, z_i), then we resort to *sampling-based PRA*, if we have been able to assign a joint probability distribution $p[x, z]$, then we may use *distribution-based PRA*, and if we have a model f such that $z = f(x)$, then we can use *model-based PRA*. Let's examine these three methods in detail.

18.5.1 Sampling-Based PRA

Say that we have a sample of n observations (x_i, z_i) and want to carry out a single-threshold PRA. Two examples are shown in Fig. 18.2. We shall be using these datasets a few times in this chapter and refer to them as the Linear and Nonlinear dataset. Our first step is to define the hazardous region H by choosing a threshold

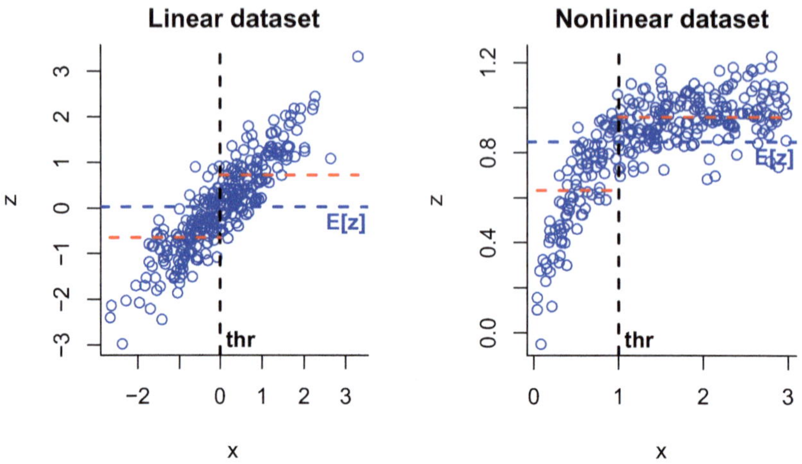

Fig. 18.2 Linear and nonlinear datasets for sampling-based single-threshold PRA. Dashed lines indicate threshold (black), overall mean (blue), conditional expectations (red)

thr, as indicated in the figure. For the Linear dataset, we choose $thr = 0$ and for the Nonlinear one we take $thr = 1$. The choice of threshold determines how the data are split into n_H 'hazardous' points on the left and $n_{\neg H}$ 'non-hazardous' ones on the right, such that $n = n_H + n_{\neg H}$. We can then without bias estimate the required expectations for z by taking averages:

$$\begin{aligned}
\hat{E}[z] &= \overline{z}, \\
\hat{E}[z|x < thr] &= \overline{z_H}, \\
\hat{E}[z|x \geq thr] &= \overline{z_{\neg H}}.
\end{aligned} \tag{18.7}$$

These expectations are indicated in Fig. 18.2. The overall expectation $E[z]$ is indicated by the dashed blue line, and the conditional expectations $E[z|x < thr]$ and $E[z|x \geq thr]$ by the shorter red dashed lines. We are now ready for the PRA. An unbiased estimate for $p[H]$ is given by n_H/n. And the PRA is completed by plugging the expectations of Eq. (18.7) into the definitions of V and R provided in Eq. (18.2). We implement this single-threshold sample-based PRA as an R-function:

```
PRA <- function( x, z, thr=0 ) {
  n  <- length(z) ; H    <- which(x < thr) ; nH      <- length(H)
  Ez <- mean( z ) ; Ez_H <- mean( z[H] )   ; Ez_notH <- mean( z[-H] )
  pH <- nH / n    ; V    <- Ez_notH - Ez_H ; R       <- Ez_notH - Ez
  return( c(pH=pH,V=V,R=R) ) }
```

When we apply the PRA function to the two datasets of Fig. 18.2, we get the following results:

Sample-based PRA

Dataset	p[H]	V	R
Linear	0.51 (0.03)	1.38 (0.08)	0.70 (0.06)
Nonlinear	0.34 (0.03)	0.32 (0.02)	0.11 (0.01)

The numbers between brackets are standard deviations, i.e. measures of uncertainty that we'll discuss later. We now implement multiple-threshold PRA. We begin by defining a new R-function for PRA that takes (besides a sample of x and z values) a vector of threshold-values:

```
PRAn <- function( x, z, thr=-1:1 ) {
  n   <- length(z) ; nthr <- length(thr) ; pH <- V <- R <- rep(NA,nthr)
  Ez_notH <- mean( z[ x>=thr[nthr] ] )
  for(i in 1:nthr) {
    Hi    <- if(i==1) which(x<thr[1]) else which(thr[i-1]<=x & x<thr[i])
    nHi   <- length(Hi) ; Ez_Hi <- mean( z[ Hi ] )
    pH[i] <- nHi / n     ; V[i]   <- Ez_notH - Ez_Hi ; R[i] <- pH[i] * V[i] }
  R.sum <- sum(R) ; pH.sum <- sum(pH) ; V.wsum <- R.sum / pH.sum
  return( list( pH.sum=pH.sum, V.wsum=V.wsum, R.sum=R.sum,
                pH     =pH    , V     =V     , R     =R     ) ) }
```

We apply the R-function to the Linear and Nonlinear datasets, choosing five threshold-levels in both cases, and plot the results in Fig. 18.3. The two datasets produce similar results for vulnerability: it decreases with resource availability (x). The Linear dataset has only few observations in the lowest x-intervals, so $p[H]$ is low there and that more than compensates for the high V, so most risk is associated with higher x-values. Because observations are more uniformly distributed in the Nonlinear dataset, risk tracks vulnerability more closely. We see that multiple-threshold PRA is much more informative than single-threshold PRA. But remember that the results of Fig. 18.3 can be reduced to equivalent single-threshold PRA results if we need to summarise. Just sum (weighted in the case of V) the results of the different x-intervals (see Eq. (18.4)).

18.5.2 Distribution-Based PRA

Let's now assume that we have the joint distribution $p[x, z]$ available to us in closed form. The distribution may have been fit to data by means of density estimation or copula selection. (For Bayesian perspectives on these two methods, see Shen et al., 2013 resp. Huard et al., 2006.) The joint distribution, together with a choice of threshold, fully determines the conditional expectations and therefore the PRA.

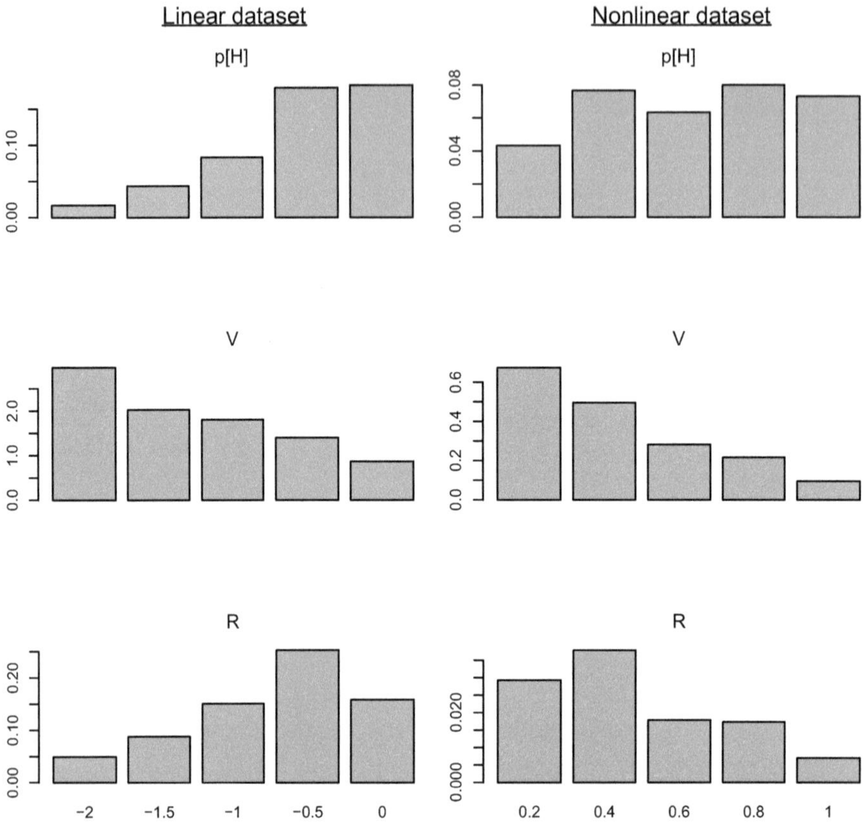

Fig. 18.3 Multiple-threshold PRA on two datasets. Five hazardous intervals were distinguished; bar labels are interval upper bounds for the environmental variable x

We do not show the details of the derivations here (but see Van Oijen & Brewer, 2022) and just present the relevant equations:

Bivariate Distributions

$$E[z|x < thr] = \frac{1}{F_x[thr]} \int_{x=-\infty}^{thr} p[x]\, E[z|x]\, dx, \tag{18.8}$$

$$E[z|x \geq thr] = \frac{1}{1 - F_x[thr]} \int_{x=thr}^{\infty} p[x]\, E[z|x]\, dx,$$

where $F_x[thr]$ (the cumulative distribution function evaluated at $x = thr$) is shorthand for $p[x < thr]$. The downside of these equations is that for nearly all probability distributions, the integrals cannot be analytically solved. They can always be solved using numerical integration, but that effectively brings us back to sampling. Sampling can be avoided in the special case of the bivariate Gaussian, for which approximations to the analytical solutions exist. Mee and Owen (1983) showed for the *standard* bivariate Gaussian (zero means, unit variances, correlation coefficient ρ) that $E[z|x < thr]$ is approximately equal to $-\rho\, p_x[thr]/F_x[thr]$. Their approximation can be generalised to non-standard bivariate Gaussians, and to $E[z|x \geq thr]$ (Van Oijen & Brewer, 2022):

Bivariate Gaussian

$$\begin{aligned} E[z|x < thr] &\approx E[z] - \rho\, \sigma_x\, \sigma_z\, \frac{p_x[thr]}{F_x[thr]}, \\ E[z|x \geq thr] &\approx E[z] + \rho\, \sigma_x\, \sigma_z\, \frac{p_x[thr]}{1 - F_x[thr]}. \end{aligned} \tag{18.9}$$

Equation (18.9) allows for a very simple approximate PRA for the bivariate Gaussian:

Approximate PRA for Bivariate Gaussian

$$\begin{aligned} p[H] &= F_x[thr], \\ V &\approx \frac{\rho\, \sigma_x\, \sigma_z\, p_x[thr]}{F_x[thr]\,(1 - F_x[thr])}, \\ R &\approx p[H]V. \end{aligned} \tag{18.10}$$

We implement this approximate PRA in R as a function of data (x, z):

```
PRAbiGauss <- function( x, z, thr=0 ) {
  mx     <- mean(x) ; sx <- sd(x) ; sz <- sd(z) ; rho <- cor(x,z)
  px.thr <- dnorm(thr,mx,sx)        ; Fx.thr <- pnorm(thr,mx,sx)
  pH     <- Fx.thr
  V      <- rho * sx * sz * px.thr / Fx.thr / (1-Fx.thr)
  R      <- pH * V
  return( c( pH=pH, V=V, R=R ) ) }
```

Note that this implementation is naive and does not account for uncertainty. It simply assumes that the mean vector and covariance matrix of the distribution equal those of the sample. But we work with this for now and leave discussion of uncertainty quantification to the next section. When we apply our R-function to the datasets of Fig. 18.2, we get the following results:

Bivariate Gaussian-based PRA

Dataset	p[H]	V	R
Linear	0.50	1.39	0.69
Nonlinear	0.28	0.27	0.08

Despite the extra step of fitting a distribution, the results are fairly similar to the direct sampling-based PRA of the preceding section. This is even the case for the Nonlinear dataset, to which a bivariate Gaussian cannot be fit well. So this naive approach is fairly robust and can be used as a quick first risk analysis before embarking on proper Bayesian density estimation and PRA with uncertainty quantification.

18.5.3 Model-Based PRA

Now we assume that we have a model f that predicts system behaviour as a function of the environment. The model can be a simple linear or nonlinear model, a complicated machine-learned model or a PBM. In all cases we can write $z = f(x|\theta) + \epsilon$. Having such a model makes it easy to derive $E[z|x]$, but we must acknowledge parameter uncertainties, at the very least for θ, and possibly also for the parameters of the error model unless we consider those known. To keep notation simple, assume that any error model parameters that need to be estimated are included in θ. In that case, Bayesian calibration on data y will give us the required $p[\theta|y]$.

To derive $E[z|x]$, these posterior parameter uncertainties must be integrated out, giving us $E[z|x] = \int f(x, \theta)\, p[\theta|y]\, d\theta$. If the model is very simple (e.g. a straight line with zero-mean Gaussian error of known variance), then the integral may be available in closed form. We then end up in the same situation as in the preceding section on distribution-based PRA for the bivariate Gaussian. But in most cases, we will again have to resort to parameter sampling to estimate the integral. Fortunately, if we have carried out the Bayesian calibration using MCMC, we will have a sample of parameter vectors, each of which we can apply (through f) to a representative set of x-values to estimate $E[z|x < thr]$ and $E[z|x \geq thr]$. The PRA then follows as earlier.

18.6 Uncertainty Quantification

The quality of our risk analysis will depend on the availability of data. If all we have is a small sample of n observations on x and z, then our estimates of $p[H]$, V and R will be quite uncertain. In the case of sampling-based PRA, we can estimate the uncertainty using standard techniques that have been developed for sampling.

Let's begin with $p[H]$. If we assume temporal independence, then hazard occurrence follows a binomial distribution $p[n_H] = Bi[n, p[H]]$. If we start from ignorance about hazard probability, then the prior for the parameter $p[H]$ can be written as a uniform distribution $U[0, 1]$ which equates to a beta-distribution $Be[1, 1]$. How does that prior change when our data arrive, i.e. when we see that n_H observations from our sample with size n are in the hazardous domain? Well, the beta prior and binomial likelihood are *conjugate* distributions (see Chap. 5 for an introduction to conjugacy), so there is an analytical solution: $p[H]$ has the following posterior beta distribution and variance:

Posterior uncertainty for $p[H]$

$$\begin{aligned} p[H] &\sim Be[a, b], \\ \sigma^2_{p[H]} &= \frac{ab}{(a+b)^2(a+b+1)}, \end{aligned} \tag{18.11}$$

where $a = 1 + n_H$ and $b = 1 + n - n_H$. The posterior variance $\sigma^2_{p[H]}$ represents our uncertainty about the hazard probability.

Let's now consider our uncertainties about vulnerability and risk. Both terms are defined as the difference between two expectations, so we consider these first. If the expectations were estimated by sampling (Eq. (18.7)), then their uncertainty can be quantified as the variances of the sample means (denoted as σ^2). These variances are estimated without bias as sample variances (denoted as Var) divided by the number of observations:

Variances of sample means

$$\begin{aligned} \sigma^2_{\hat{E}[z]} &= \frac{Var[z]}{n}, \\ \sigma^2_{\hat{E}[z|x<thr]} &= \frac{Var[z_H]}{n_H}, \\ \sigma^2_{\hat{E}[z|x\geq thr]} &= \frac{Var[z_{\neg H}]}{n - n_H}. \end{aligned} \tag{18.12}$$

The variance of the difference between two independent terms is the sum of the two variances. The two terms that define V are independent because $x \geq thr$ does not overlap with $x < thr$. So the uncertainty for V is simply the sum of the last two variances of Eq. (18.12). But the two expectations that define R are not independent, and we need to correct for their covariance. This leads to the following uncertainty quantifications for V and R (Van Oijen & Brewer, 2022):

Posterior uncertainty for $V$and R

$$\begin{aligned} \sigma^2_V &= \sigma^2_{\hat{E}[z|x\geq thr]} + \sigma^2_{\hat{E}[z|x<thr]}, \\ \sigma^2_R &= \sigma^2_{\hat{E}[z|x\geq thr]} + \sigma^2_{\hat{E}[z]} - 2(1 - \frac{n_H}{n})\sigma^2_{\hat{E}[z|x\geq thr]}. \end{aligned} \tag{18.13}$$

We can extend our PRA function for PRA with this uncertainty quantification:

```
PRA.UQ <- function( x, z, thr=0 ) {
  n  <- length(z) ; H    <- which(x < thr) ; nH      <- length(H)
  Ez <- mean( z ) ; Ez_H <- mean( z[H] )   ; Ez_notH <- mean( z[-H] )
  pH <- nH / n    ; V    <- Ez_notH - Ez_H ; R       <- Ez_notH - Ez
  a           <- 1 + nH ; b <- 1 + n - nH
  s2_Ez       <- var(z    ) /  n
  s2_Ez_H     <- var(z[ H]) /    nH
  s2_Ez_notH <- var(z[-H]) / (n-nH)
  s_pH        <- sqrt( a*b / (a+b)^2 / (a+b+1) )
  s_V         <- sqrt( s2_Ez_notH + s2_Ez_H )
  s_R         <- sqrt( s2_Ez_notH + s2_Ez - 2 * s2_Ez_notH * (1-nH/n) )
  return( c( pH=pH, V=V, R=R, s_pH=s_pH, s_V=s_V, s_R=s_R ) ) }
```

The code generates values for $p[H]$, V and R together with their uncertainties expressed as standard deviations.

So far, we discussed uncertainties generated in sampling-based PRA. Let's now discuss uncertainty quantification in distribution-based PRA and model-based PRA. In these cases, sampling uncertainty need not play a role. We can generate as many virtual observations (x_i, z_i) from any parameterised distribution $p[x, z]$ or model $f(x)$ as we like, reducing sampling uncertainty to nearly zero. But parameter uncertainty remains an issue, as does uncertainty about our choice of distribution and model. Bayesian calibration and model comparison are warranted to quantify these uncertainties. The study of copulas and of extreme value theory (e.g. Serinaldi et al., 2020) has confirmed the obvious: risk analysis is very sensitive to the tail behaviour of distributions. Because hazardous conditions are often rare extreme events for which data are scarce, careful parameterisation and comparison of distributions and models remain critical.

18.7 Expanding the Theory

This chapter introduced probabilistic risk analysis theory as developed by Van Oijen & Brewer (2022). We conclude by briefly mentioning three ways to expand the theory (covered more fully in the same reference).

- Risk-decomposition can be expanded to include *exposure* (U), such that $R = p[H]\, U\, V$. This is useful when V and R—in our initial PRA—referred to scale-free intensive variables, and we want to study the extent of the impact. For example, say that V and R referred to cumulative productivity per unit area (kg m^{-2}) but that regions differ in the area U exposed to the hazard. Multiplication by U in units of area (m^2) will then convert R to an extensive variable (kg) for each region, while V remains intensive (kg m^{-2}). By including exposure, we can thus implement a spatially distributed risk analysis.

- In all examples in this chapter, hazardous conditions H were defined as an environmental variable x being below a specific threshold. It would be straightforward to replace this with x being damaging *above* a threshold (by swapping the roles of the two conditional expectations, or by a change of variable $x' = -x$). But in fact we have complete freedom in how we define H. A *categorical PRA* would be one where we define the hazardous region as being a specific *type* of environmental condition, e.g. an El Niño climate year, or a repeated (rather than first-time) occurrence of an environmental extreme or an economic recession.
- PRA can play a role in decision-making. The links between PRA and Bayesian decision theory will be examined in the next chapter.

Exercises

1. **Derivation of V.** Consider Eq. (18.1). Show how the equation for V follows from the two equations for R.
2. **Approximate PRA for bivariate Gaussian** $p[x, z]$. Derive Eq. (18.10) from Eq. (18.9).
3. **Uncertainty as a function of number of data.** How does the uncertainty of sampling-based risk analysis terms ($p[H]$, V, R) change if there are twice as many data (x, z)?

Chapter 19
Bayesian Decision Theory

19.1 Basic Equations

Risk analysis (Chap. 18) is useful to summarise the implications of predictive uncertainty for risk, but for decision-making a wider framework is required, *Bayesian decision theory* (BDT) (Berger, 1985; Jaynes, 2003; Lindley, 1991; Van Oijen & Brewer, 2022; Williams & Hooten, 2016). In BDT, every decision problem has three main ingredients:

1. A list or continuum of possible actions $a \in A$: exactly one action is to be decided upon.
2. A list or continuum of possible external conditions $x \in X$. These are uncertain, so we have $p[x]$, possibly $p[x|a]$.
3. A *utility* function $u(a, x)$ that can be evaluated for every combination of a and x.

For every decision a, the *expected utility* is defined as:

$$\overline{u}[a] = E[u(a)] = \int p[x]\, u(a, x)\, dx. \tag{19.1}$$

If we have a finite list of possible conditions x_i, then $p[x]$ is a discrete probability distribution and $\overline{u}[a]$ is written as a sum rather than an integral, but the idea remains the same. The Bayesian solution to the decision problem is simply: choose the action a^* that maximises the expected utility:

$$\overline{u}[a^*] = \max_{a \in A} \overline{u}[a]. \tag{19.2}$$

But what is Bayesian about that? Well, nothing as such except that we require probabilities and utilities to be *coherent* so that we can use the rules of probability theory, including Bayes' Theorem, to manipulate them. This is especially important if we want to break down the decision problem in a sequence of smaller decisions

M. van Oijen, *Bayesian Compendium*,
https://doi.org/10.1007/978-3-031-66085-6_19

with more easily quantified conditional probabilities, and when we are receiving new information that can help us update our probability distributions. Say that we have a decision to make and have just received relevant data y. Data are relevant if $p[x|y] \neq p[x]$. We can then identify the action $a = a_y^*$ that maximises the *posterior expected utility*:

$$\overline{u}[a_y^*] = \max_{a \in A} \int p[x|y]\, u(a, x)\, dx, \tag{19.3}$$

where we may use Bayes' Theorem and replace $p[x|y]$ with $p[x]\, p[y|x]$. The extra information may thus lead to a different optimal action. We can quantify the value of the information y as follows. Imagine that we have not received any data yet. Getting data may be expensive, so we are not sure yet whether data acquisition is worth it. For example, we may consider buying a local weather forecast y that would give us information about our field conditions x. We obviously don't know what the forecast will be—any $y \in Y$ is possible—but experience tells us that $p[y]$ describes our uncertainty well. And for every possible value y, the expected utility will be $\overline{u}[a_y^*]$ rather than $\overline{u}[a^*]$. So the expected utility after the upcoming data acquisition will be:

$$\overline{u}[a_Y^*] = \int_Y \overline{u}[a_y^*]\, p[y]\, dy. \tag{19.4}$$

The *Value of Information* is therefore:

$$VoI = \overline{u}[a_Y^*] - \overline{u}[a^*]. \tag{19.5}$$

The VoI helps us choose between deciding immediately and first acquiring more data.

19.2 A Graphical Model for Decision-Making

We began this chapter with a short list of three ingredients that every decision problem has: a, x and $u(a, x)$. Let's now add some flesh to these bare bones. The *graphical model* (GM) of Fig. 19.1 depicts a more detailed joint probability distribution for decision-making under uncertainty. (See Chap. 16 for an introduction to graphical models.) It shows utility u as the difference between the *benefit* from system performance z and the *cost* associated with action a. Our probability distribution for z is formed by a model prediction $f(a, x, \theta) + \epsilon$ that accounts for uncertainties in x, parameters θ and model error ϵ. Most decision problems fit into this scheme.

Let's begin with a simple linear example to make things more concrete. Consider Fig. 19.2. It represents the problem of deciding how much irrigation to use to

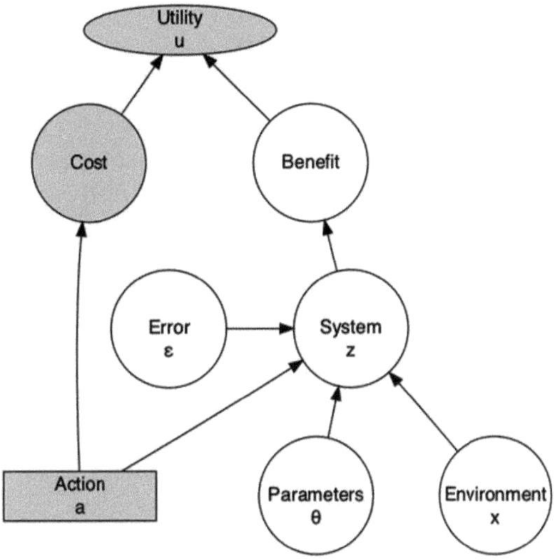

Fig. 19.1 A graphical model for Bayesian decision theory

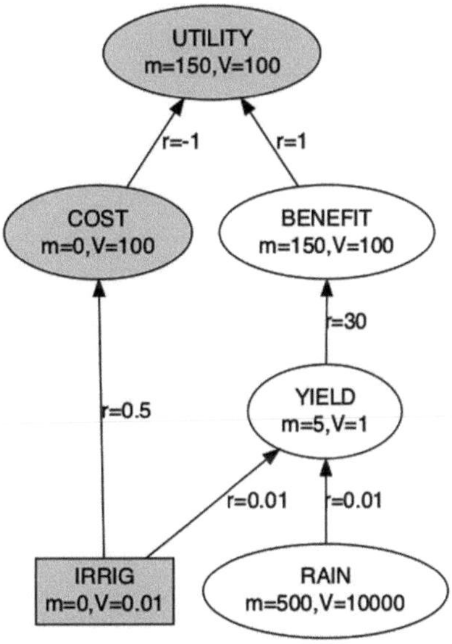

Fig. 19.2 Example of linear BDT represented by a DAG. Nodes show the values of the prior mean and conditional variance. Edges are labelled with the regression coefficient

improve crop yield. The model is a Gaussian Bayesian Network (GBN), a type of GM that we discussed at length in Chap. 16. GBNs represent multivariate Gaussian distributions, and all parameters that together define the joint probability distribution are supplied in the figure: means and variances in the nodes, and regression coefficients next to the edges. We also recognise the elements of Fig. 19.1 in the GBN. The to-be-decided-upon action a is 'IRRIG', the environment x is 'RAIN', system performance z is 'YIELD', and utility u is 'UTILITY'. You can interpret the regression coefficient next to the arrow from IRRIG to COST as the per-unit price of irrigation water, and the one next to YIELD → BENEFIT as the selling price of crop yield.

We could now use the techniques presented in Chap. 16 to calculate the expected means (and variances) for UTILITY across the range of possible values for IRRIG. But ... a GBN like this is typically not a useful probabilistic model for decision-making because it is linear, and real-world problems tend to be nonlinear. GBNs are usually oversimplifications. Moreover, most of the information in the GBN is not needed to solve our decision problem. The numbers in the graph immediately tell us which irrigation level is optimal, but we leave this as an exercise to the reader.

19.3 A Nonlinear Example

We now turn to a more interesting nonlinear example. Let system performance be a negative exponential performance function: $z = f(a, x, \theta) + \epsilon = \theta(1 - e^{-a-x}) + \epsilon$. Further, let cost remain proportional to action level and benefit proportional to system performance. We implement the resulting utility function in R and assign some default values to environment and parameters:

```
u <- function( a, x=1, t=1, e=0, ka=0.2, kz=1 ) {
  z       <- t * (1-exp(-a-x)) + e
  cost    <- ka * a
  benefit <- kz * z
  return( benefit - cost ) }
```

If we were certain that our default settings for all variables and parameters were correct, then the decision problem would be a simple matter of identifying the value of action a that maximises the utility, as in the left panel of Fig. 19.4. We find that without uncertainties, $a^* = 0.6$. But we do have uncertainties, so the Bayesian decision problem is to find the value of a that maximises $\overline{u}[a] = E[u(a)]$ which we must calculate by integrating out the uncertainties. Say that our uncertainties are represented by the following probability distributions:

$$\begin{aligned} x &\sim N[1, 1], \\ t &\sim N[1, 0.5], \\ e &\sim N[0, 1], \\ ka &\sim U[0.1, 0.3], \\ kz &\sim U[0.5, 1.5]. \end{aligned} \tag{19.6}$$

To illustrate the impact of these uncertainties for x and parameters on utility predictions, we can generate a large sample of vectors and calculate the utility with the R-function that we just defined. We show the results for $a = 0$ and $a = 2$ in

Fig. 19.3. The higher value of a reduces the correlation between x and u because of the asymptotic behaviour of the negative exponential system function $(1 - e^{-a-x})$.

The analysis of this decision problem with uncertainty requires that we repeat this sampling exercise with values of a over its whole domain, which we choose to be $A = [0, 0.1, .., 1.9, 2]$. For each $a \in A$, we calculate the expected utility from the sample. The results are shown in the middle panel of Fig. 19.4. We see that the optimum action is now 1.1, which is larger than it was without uncertainty. And if we increase the uncertainty by 50%, then the optimum action increases even further to 1.7 (right panel of Fig. 19.4).

We conclude that the degree of uncertainty may affect the value of the optimum action. This is a common phenomenon in decision-making. For an amusing example, see Tibshirani et al. (2011). They showed that the optimum target for a

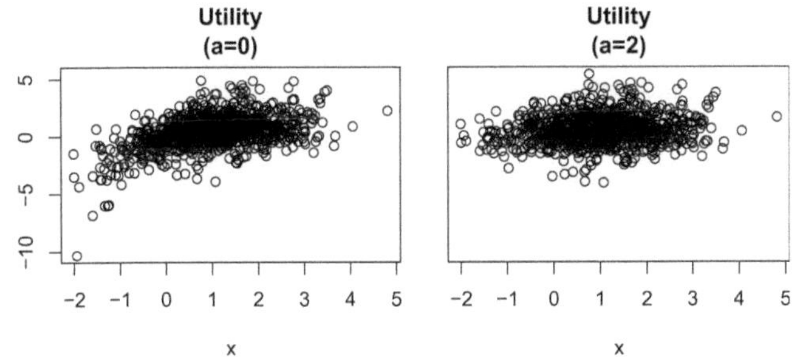

Fig. 19.3 Utility as a function of environment x for a negative exponential performance function and linear cost and benefit functions. Left: action $a = 0$; right $a = 2$. Both panels are for the same sample (n = 1000) of x and parameters

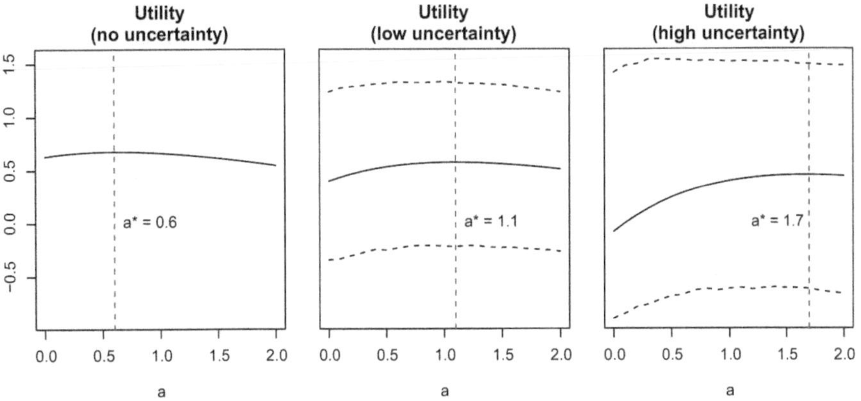

Fig. 19.4 Expected utility as a function of action a for a negative exponential performance function and linear cost and benefit functions. Left panel: no parameter uncertainty. Middle panel: default uncertainties (see text). Right panel: uncertainties multiplied by 1.5. Solid line: expectation. Dashed lines: Q25 and Q75

darts-thrower to aim at depends on the precision with which they can throw the darts.

19.4 PRA and BDT

How does the probabilistic risk analysis (PRA) that we developed in Chap. 18 relate to the BDT? Is maximisation of utility expectation somehow equivalent to minimising risk? That is the topic of this section. For discussion of the links between older (discrete and less comprehensively probabilistic) forms of PRA and BDT, see Borgonovo et al. (2018).

To compare the PRA and the BDT, they should refer to the same quantities. So we choose utility u as the PRA's system variable. But what about the PRA's environment variable? Let's focus on the nonlinear decision problem of this chapter. We could use the BDT environment variable x also as the PRA environment variable. But in the BDT example, x and a together drive the performance of the system. The negative exponential performance function defines a dependence of the system on $x + a$, so we choose that sum as the environment variable for our 'BDT-mimicking' PRA. We are free to choose any threshold value (it is not part of BDT); we choose $thr = 0$.

The idea is now to vary the action a and see how it affects the risk. We expect that increasing the action a will decrease the frequency with which our PRA environment variable $(x + a)$ drops below the threshold. In other words, this is an example of an action that will decrease $p[H]$ but may have little impact on V. R will then primarily be reduced because of the effect of the action a on $p[H]$ rather than on V. We will run a separate PRA for each value of the action $a \in A$ that we examined in the BDT, and we again compare results for different levels of uncertainty. We use the code for PRA that we developed in Chap. 18. The results are shown in Fig. 19.5 and partly confirm our expectations. We see that $p[H]$ indeed changes the most when varying a, but V also changes. We further note that high uncertainty leads to higher risk.

The PRAs that we just carried out for our BDT example show that we can make risks as small as we want by increasing the action a. But remember that the risk is the *difference* between two expectations for the system variable of interest (here the utility u), and both expectations include the same cost of the action. In other words, increasing a may reduce R to zero but that may not be desirable because the action costs $ka \times a$ will increase and thereby decrease u itself. Another way of looking at this is that a mitigating action may make a hazard factor (here x) irrelevant and thus stabilise system performance at the cost of reducing utility even in situations where the hazard factor would not have been a problem anyway. Note that this is how paying for insurance works!

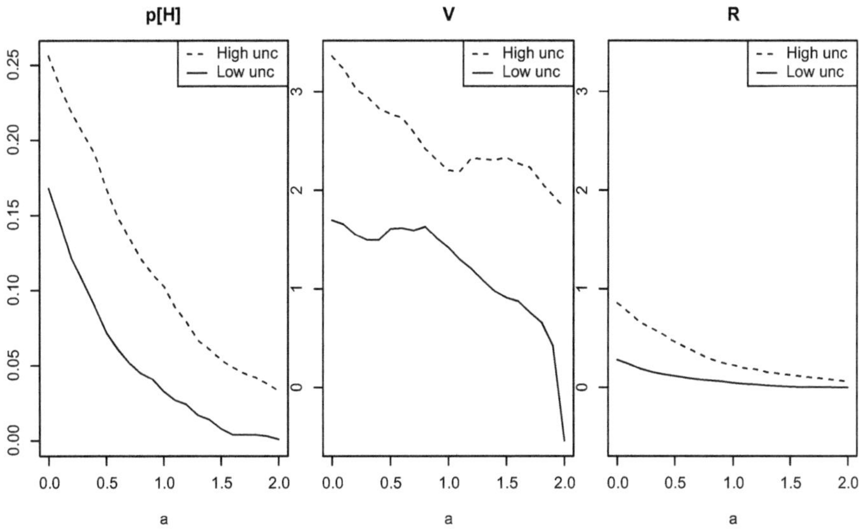

Fig. 19.5 PRAs on data generated for BDT

Let's now examine the relationship between risk R and utility u more formally. Because u is our system variable of interest, R is here defined as:

$$R = E[u|\neg H] - E[u]. \tag{19.7}$$

So maximising $E[u]$ is the same as minimising $R - E[u|\neg H]$. We interpret the latter as 'risk corrected for costs and benefits' and denote it as R_c:

$$\begin{aligned} R_c &= R - E[u|\neg H] \\ &= R + E[ka|\neg H] \times a - E[kz \times z|\neg H]. \end{aligned} \tag{19.8}$$

Let's now verify that the corrected risk is indeed minimised at the same values of the action a that maximise expected utility. Figure 19.6 shows the results for the same two levels of uncertainty that we explored above, and we see that the smallest values of R_c are indeed found for the same values of a as maximising $E[u]$ did before (Fig. 19.4). So BDT can be formulated in two equivalent ways: as maximising expected utility $E[u]$ and as minimising corrected risk R_c.

Our intention in defining R_c was to clarify the relationship between risk analysis and decision-making. But note that R_c cannot replace R completely. Only R can be written as the product of $p[H]$ and V; R_c cannot be decomposed in that way. So there remains a useful role for R, and risk analysis in general, in helping us understand our systems and the consequences of our actions.

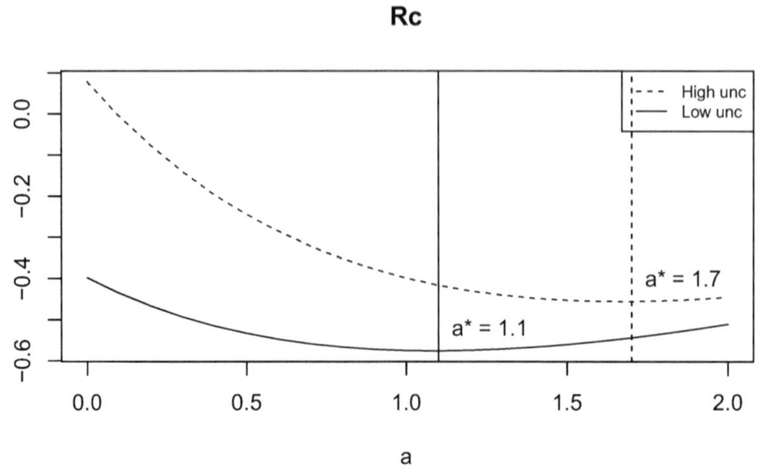

Fig. 19.6 Impact of action a on corrected risk R_c calculated from data generated for BDT. R_c reaches an uncertainty-dependent minimum at $a*$

Finally, when looking back at this chapter and the preceding one, we see that risk analysis and decision theory bring together many of the themes of this book: modelling, data acquisition, uncertainty assessment, Bayesian calibration, graphical modelling. PRA and BDT are tools for making models useful in practice.

Exercises

1. **Linear BDT.** Consider the GBN depicted in Fig. 19.2. Why is this linear decision problem trivial? Deduce the optimal amount of irrigation directly from the graph, without any computer assistance.
2. **Life expectancy.** This is a binary decision problem: the action is either to take a medicine or not. A dangerous epidemic is coming and everyone's chance of getting the disease is $p[H] = 0.5$. If you take no medicine ($\neg M$) and stay disease-free, your remaining life expectancy is 100 minus your age x in years: $E[z|\neg H, \neg M] = 100 - x$, but without medicine the disease is lethal: $E[z|H, \neg M] = 0$. With medicine, your life expectancy is always $E[z|H, M] = E[z|\neg H, M] = 100 - x - x^2/100$. The benefit of the medicine is the expected increase in z when the disease hits you, and the cost is the expected loss of z due to medicine side effects when the disease passes you by. Below what age should you be taking the medicine? (You can inspect for yourself how parameter uncertainties would change the outcome.)

Chapter 20
Linear Modelling: LM, GLM, GAM and Mixed Models

Science increasingly recognises the nonlinearities in nature, and Bayesian methods can handle nonlinear models without any problem. However, linear modelling remains the default statistical approach for many, and it is therefore important to be familiar with the field. This chapter gives a brief overview of the most common classes of linear modelling.

20.1 Linear Models

Linear models are models whose output depends linearly on parameters. So the quadratic model $f(x, \theta) = \theta x^2$ is a linear model even though the dependence on x is not linear. More complicated models like $f(x, \theta) = \theta_1 x_1^2 + \theta_2\ log(x_2 x_3)$ are also linear in their parameters. That shows that quite a wide variety of system behaviour can be represented by linear models. However, linear models are models of convenience. We do not use them because nature is linear, but because they are well-behaved and make data analysis easy. So they can be very useful in a first exploration of new data, but we should be careful when using them for prediction, especially when extrapolating the predictor variables x beyond values for which we have observations.

Before introducing different classes of linear modelling, we need to revisit our earlier discussion of the meaning of the term *model* (Chap. 1). Generally, in this book, we define a model as a function f of predictor variables x and parameters θ. But when we expand the model with a term that quantifies our uncertainty about measurement or modelling error, then we must distinguish between the *core model* f and the *statistical model* $f + \epsilon$ (with $\epsilon \sim p[\epsilon]$). When people apply linear modelling in their work, they generally use the term 'model' to indicate the whole statistical model, and they refer to the core-model part as the 'linear predictor'. In short, a linear model is a statistical model that is written as the sum of a linear

M. van Oijen, *Bayesian Compendium*,
https://doi.org/10.1007/978-3-031-66085-6_20

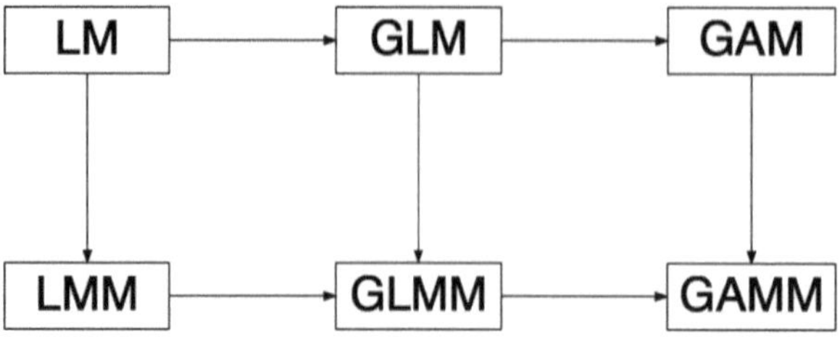

Fig. 20.1 Generalisation of the basic linear model (LM) to generalised linear and additive models (GLM, GAM) and mixed models (LMM, GLMM, GAMM)

predictor f and an error term ϵ with an appropriate probability distribution assigned to the latter. For ease of discussion in this chapter, we shall consider the differences between a correctly parameterised linear predictor and the observations y to be due to measurement error (not model error), so we can write $y = f(x, \theta) + \epsilon_y$.

We shall restrict our discussion of linear modelling to the family tree depicted in Fig. 20.1. The tree shows the six most common classes of linear statistical models, from simple LM to more complex and flexible GLM and GAM—and their extensions to 'mixed models'. In all classes, the assumption is that the mean value of the distribution $p[\epsilon]$ is zero, so the expectation value for as yet unobserved data y is $E[y] = f(x, \theta)$.

20.2 LM

Linear statistical models (LM) are the first models that we consider here. In LM, the linear predictor is a simple regression on the predictor variable(s), and the error term has a zero-mean Gaussian distribution: $f(x, \theta) = X\beta$ and $\epsilon_y \sim N[0, \Sigma_y]$. We have already seen examples of LM in this book, in the form of linear regression using OLS, WLS and GLS (Chap. 14). These are examples of LM that only differ in the Σ_y term.

20.3 GLM

Generalised Linear Models (GLM) differ from LM in both the linear predictor and in the error term. In GLM, the linear predictor is wrapped in a transformation function and the error term need not be Gaussian but can be another distribution in the *exponential family of probability distributions* which includes, besides the Gaussian, distributions such as the Poisson, binomial and gamma. The transformation is generally applied to $E[y]$ rather than to f, so we have $g(E[y]) = X\beta$, where $g()$ is called the *link function*. And the error distribution is usually only implicitly defined through a data distribution $p[y]$. (We can of course retrieve our more familiar model expression by writing $y = E[y] + \epsilon_y$ where $E[y] = g^{-1}(X\beta)$ and $\epsilon_y \sim p[y] - E[y]$.)

Common examples of GLM are *logistic regression*, which uses a logit link function and a binomial distribution, and *Poisson regression*, which uses a log link function and a Poisson distribution. Note that if you choose the identity link function ($g(E[y]) = E[y] = X\beta$) and a Gaussian distribution, the GLM reverts to a simple LM because $y \sim N[E[y], \Sigma_y]$ is then equivalent to $y = X\beta + \epsilon$ with $\epsilon \sim N[0, \Sigma_y]$.

20.4 GAM

Generalised Additive Models (GAMs) also use link functions, but they define $g(E[y])$ not as a linear regression $X\beta$ but as a sum of smooth functions $s(x)$ or a mixture of one or more regression terms plus one or more smooth functions. The 'smooths' can be functions of individual predictor variables or multiple variables, so statistical interactions can be represented. Typically, each $s(x)$ is written as the weighted sum of several *basis functions* which are often *splines*. So a GAM has many parameters that need to be estimated. But the fact that a GAM is still a linear model with respect to the basis function weights facilitates model fitting.

20.5 Mixed Models

Mixed models are hierarchical models. This means that at least for some of their parameters we not only estimate the values of the parameters themselves, but also properties ('hyperparameters') of the population that they belong to. A simple example would be a varying-intercept, common-slope model $y_{ij} = \theta_j + x_{ij}\beta + \epsilon_{ij}$, with $\theta_j \sim N[0, \sigma_\theta^2]$ for $i \in \{1..n_j\}$, $j \in \{1..J\}$. This is a linear mixed model (LMM) in which the hyperparameter σ_θ^2 represents the variance of the intercept between groups. (The population mean of θ is assumed to be zero in this example, but it can be linearly related to covariates whose values differ between groups.) In typical confusing linear modelling jargon, the θ_j would be referred to as 'random effects' and the β as a 'fixed effect', but of course there is nothing random about the former and nothing fixed about the latter. There are just different degrees of, and representations of, variability. In practical use of mixed models, random effects are often nuisance parameters that people are not interested in (so a summary in terms of their variance is enough), whereas the fixed effects quantify the impact of predictor variables that are of primary interest to the researcher. You may for example be interested in the impact of a treatment variable like resource availability, but not in how differences between experimental blocks, locations or times affect that impact. By treating the latter differences hierarchically, you quantify their variability. You can then account for that variability as a known source of uncertainty when making predictions of the treatment effect on new members of the same population.

Similar extensions to hierarchical modelling can of course also be applied to GLM and GAM, which then become GLMM and GAMM (e.g. Diggle & Ribeiro, 2007; Gelman et al., 2013; Wood, 2006).

20.6 Parameter Estimation

The main motivation for linear modelling is its simplicity and concomitant ease of parameter estimation. Most studies that use linear modelling estimate parameter values by means of (restricted) maximum likelihood estimation rather than full Bayesian calibration. For the simple LM, there are analytical solutions (Lindley & Smith, 1972), as we saw in Chap. 14, but when we move to the more flexible models in our family tree, the methods for parameter inference become increasingly approximate, especially with small datasets (Wood, 2006). The flexible models also run the risk of following the data too closely, making the response surfaces very wiggly. To avoid such tracking of noise rather than signal, we may want to introduce a 'penalty' to high curviness of the function that we want to fit, e.g. by reducing the likelihood value by the average second derivative of the function. In such *penalised regression*, maximum likelihood estimation is replaced by identifying the maximum of (something akin to) likelihood minus penalty. From a Bayesian perspective, the penalty function constitutes a prior on function shape. A more principled approach is to go full Bayesian and ensure that prior and likelihood are expressed as proper probability distributions. But if we do so, the beautiful theory developed for linear modelling will not help us much anymore: we will quickly need to resort to MCMC. For the mixed models, which are hierarchical, Gibbs sampling or other MCMC algorithms that include a Gibbs-step, will be appropriate.

20.6.1 Software

Models in the class of LM can be fit analytically, but maximum-likelihood parameter estimation of LM and GLM can also be done with the basic R-functions `lm` and `glm`. For LMM and GLMM, there is the R-package `lme4`, and for GAM and GAMM you can use the package `mgcv`, with `gamm4` as a more recent alternative. But none of these packages allows you to specify a prior for the parameters, they just carry out maximum likelihood estimation. So for a Bayesian approach, we have to look elsewhere.

R-package `arm` allows you to run a GLM with some control over the prior. More powerful are the packages `rstanarm`, and `brms` which can run all the model types discussed in this chapter. Both these packages make use, behind the scenes, of the Hamiltonian MCMC engine built into the general-purpose statistical software

Stan. For GLMM, there is also `MCMCglmm`. You can of course also do more of the programming yourself. In Chap. 17, we calibrated a LMM using the R-package `rjags` that runs MCMC in JAGS. JAGS gives you more freedom to choose priors and likelihoods than the aforementioned packages do.

To give you a flavour of some of the R-packages, consider the following lines of code.

```
lm.1 <- lm       ( y~x )
lm.2 <- glm      ( y~x, family=gaussian(link=identity) )
lm.3 <- bayesglm( y~x, family=gaussian(link=identity) )
lm.4 <- brm      ( y~x, family=gaussian(link=identity), data=xy )
lm.5 <- stan_glm( y~x, family=gaussian(link=identity), data=xy )
lm.6 <- gam      ( y~x, family=gaussian(link=identity) )
```

The first two commands are base R, and the next four are from R-packages `arm`, `brms`, `rstanarm`, and `mgcv`. All six fit the same straight line, as you will have expected given the use of the Gaussian distribution family and identity link function. The fitted line is shown in the leftmost panel of Fig. 20.2. More interesting application examples are produced by the following three commands which produce the results shown in the other panels of the figure.

```
lmm.1 <- brm      ( y~x+(1|group), data=xgy )
glm.1 <- stan_glm( y~x, family=gaussian(link=log), data=xy )
gam.1 <- gam      ( y~s(x) )
```

Default settings were used for function controls and parameters. We leave further study of the various models and packages to the reader.

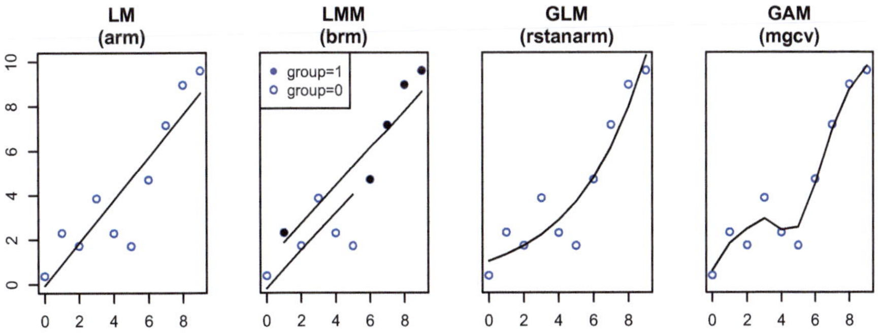

Fig. 20.2 LM, LMM with varying intercept and common slope, GLM with log link-function, GAM with thin plate splines. Each model is applied to the same dataset (x,y). In brackets: software used

Exercise

1. **Explore the R-packages.** Explore the two most powerful Bayesian packages mentioned in this chapter: `brms`, and `rstanarm`. What are their default prior probability distributions? Vary those priors and see how they affect the results shown in the middle two panels of Fig. 20.2. Add measures of output uncertainty to the figure. Plot the posterior parameter distributions.

Chapter 21
Machine Learning

Machine learning is the name for a very wide collection of techniques for exploring data and estimating functions, and we can only scratch the surface here. The field has expanded and diverged to the extent that it is hard to find common denominators. But we can say that most machine learning techniques do not focus on explanation or on uncertainty quantification, but on prediction. Machine learning is typically applied in cases where we have many data but not much understanding of causal pathways to guide our modelling.

The field has developed with more input from computer scientists than from statisticians or natural scientists. This may be surprising because machine learning is the offshoot from artificial intelligence research that aimed to build process-based models that mimic the behaviour of the human brain, with *neural network models* as the classical example (McCulloch & Pitts, 1943). Despite this early aim of understanding how humans learn, machine learning now mostly consists of black-box approaches whereby the machine—rather than the human—learns to develop predictive algorithms from data. This has given us a problem with interpretability (Hanson et al., 2023). You can open the black box model, but when you peek inside, you won't find anything resembling real-world processes. The algorithms developed by means of machine learning can be so complicated that we now see researchers in their turn developing computer algorithms for interpreting the computer-generated algorithms! An important motivation for developing these interpretation techniques is that machine learning algorithms are increasingly used in human society and need to be challenged (Nature Editorial, 2023; O'Neil, 2016). The machine learning interpretation algorithms that have been developed so far are predominantly different forms of sensitivity analysis that help us identify the main predictor variables and interactions (Molnar, 2023). It remains hard to derive ideas about causal pathways from these algorithms (Rudin et al., 2022).

Besides the problem of poor interpretability, the second main problem of machine learning is that the methods are not geared towards uncertainty quantification. Machine learning uses training data to maximise the predictive capacity of

M. van Oijen, *Bayesian Compendium*,
https://doi.org/10.1007/978-3-031-66085-6_21

highly flexible functions, trying to find unique optimal parameter vectors rather than posterior probability distributions. It has long been recognised that there is much scope for Bayesian methods to improve on this way of working (Bishop, 2006; MacKay, 1995; Murphy, 2023, 2022), but progress has been quite limited. Of course, Bayesian machine learning is nothing special—assign a prior and likelihood and off you go—but finding the posterior distribution is often computationally demanding.

21.1 The Family Tree of Machine Learning Approaches

Machine learning methods are generally classified according to the family tree of Fig. 21.1. The first division is between *supervised* and *unsupervised* learning (Murphy, 2022). The terms are perhaps misleading—human intervention is always necessary—but they refer to the nature of the data that the methods operate on. In the case of supervised learning, the training data are (x, y), i.e. both predictor and dependent variables, whereas unsupervised methods only explore independent variables x. Supervised learning derives algorithms for predicting y, whereas unsupervised learning identifies patterns in x. The supervised methods are subdivided further into methods for *regression* and *classification*. The only difference is that in regression the dependent variable y is continuous, whereas in classification it is discrete and can be associated with class labels. Various machine learning methods can be used for both purposes. Unsupervised methods are subdivided into methods for *clustering* the data x into different groups, and methods for *dimensionality reduction* which find low-dimension projections of x that preserve most of the information in the data.

We can view the four classes of machine learning as attempts to find optimal functions $f(x, \theta)$ for different purposes. The classes differ in the nature of the training data and in the constraints that they put on the shape of $f()$. In supervised learning, we use flexible functions $f()$ such that they can reproduce any kind of

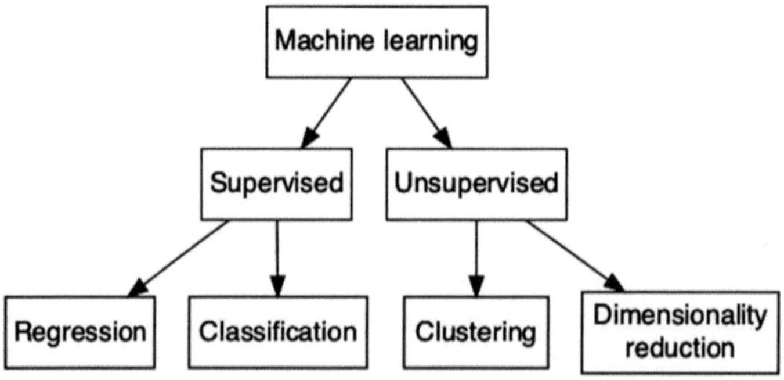

Fig. 21.1 Family tree of machine learning methods

pattern existing in the training data (x, y). Typical examples are neural networks for regression, and *decision trees* (Chipman et al., 2010) and *random forests* (ensembles of decision trees), for classification. We shall discuss neural networks in more detail below to illustrate some of the key strengths and weaknesses of machine learning. Easy-to-use but less flexible supervised classification methods are *logistic regression* (an example of GLM, see Chap. 20), *support vector machines* and the *k-nearest neighbour* algorithm. Many of the mentioned classification methods can be used for regression and vice versa (James et al., 2013).

In unsupervised clustering, we try to find a function $f(x, \theta)$ that assigns labels to the x such that distances between cluster members are smaller than distances between clusters, with many different options for the choice of distance function. Examples are probabilistic *Gaussian mixture modelling* and its deterministic simplification *k-means clustering*. Finally, the main example of dimensionality reduction is *principal component analysis* (PCA) where we sequentially identify orthogonal linear data vector transformations ('components') $f_i(x, \theta_i) = x\theta_i$ such that θ_i is a unit vector and $Var[f_i]$ is maximised. By keeping only the first few components, which together account for most of the variance in the data, we reduce the dimensionality of x.

In practice, the role of the unsupervised learning techniques is mainly in the initial phase of studies, when we want to reduce the complexity of our data. The clusters and the transformed data of lower dimension can then be passed on to the supervised methods.

21.2 Neural Networks

Neural networks are functions. They take uni- or multivariate x as input and produce uni- or multivariate output. They are generalisations of regression models such that a function of any shape can be approximated to any desired precision with a neural network. An example of a simple neural network is shown in Fig. 21.2. In

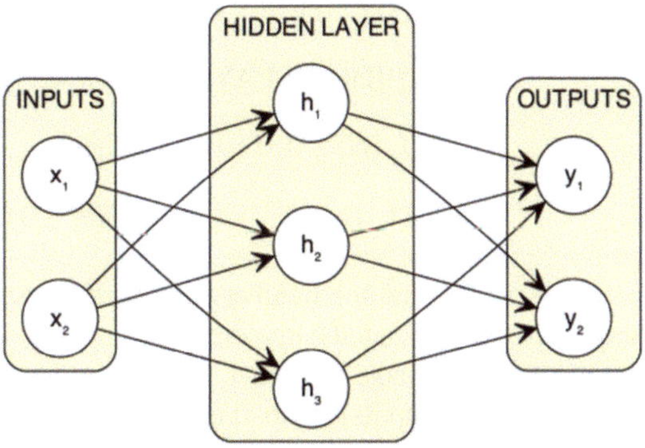

Fig. 21.2 A 2-3-2 neural network. Each neuron is connected to all inputs and to all outputs and there are no loops, making this a standard feedforward neural network

the neural network of the figure, there is a layer of network nodes between the inputs and outputs—we say that there is one 'hidden layer' of neurons. In *deep learning*, more layers will be present. Let's first study one-hidden-layer networks, which will already turn out to be remarkably flexible functions, before showing the consequences of moving to more layers.

21.2.1 Neural Networks with One Hidden Layer

Figure 21.2 shows a 2-3-2 network: it has two inputs, three neurons in a single hidden layer, and two outputs. There is great modelling freedom here, everything upward from 1-1-1 is possible. But what function does such a network represent, and what happens at the neurons? Well, each neuron is a small model in its own right that (1) calculates a weighted sum of its inputs, (2) adds a 'bias', (3) transforms the result. In other words, each neuron is a generalised linear model (GLM; see Chap. 20). The transformation function is, in analogy to neuron activity in the brain, called an *activation function*. Activation functions are generally chosen to be nonlinear and often S-shaped such as the logistic or the hyperbolic tangent (written as `tanh()` in R). If you choose the former, your neurons will be mimicking logistic regression. If we assume that the latter function was used in our network, then the mini-model represented by the neuron labelled as h_1 is:

$$\begin{aligned} h_1 &= \tanh(b_1 + w_{11}x_1 + w_{12}x_2) \\ &= \tanh(b_1 + \Sigma w_{1i}x_i) \\ &= \tanh(X\theta_h^{(1)}), \end{aligned} \tag{21.1}$$

where $X = (1, x_1, x_2)$ and $\theta_h^{(1)} = (b_1, w_{11}, w_{12})^\top$. Let's now turn to the output layer. The output nodes are the same kind of mini-models as the hidden-layer neurons, but they have their own weights, biases, and activation function. Taking everything together, we can write the function expressed by neural networks like this one very compactly as follows:

<u>ONE-HIDDEN-LAYER NEURAL NETWORKS:</u>

$$\begin{aligned} y_i &= f_y(H\theta_y^{(i)}),\ i = 1..n_y \text{ where} \\ H &= (1, f_h(X\theta_h^{(1)}), \ldots, f_h(X\theta_h^{(n_h)})), \end{aligned} \tag{21.2}$$

where n_h and n_y are the numbers of nodes in the hidden and output layers, and θ_h and θ_y are the associated parameters. The main point of writing the network function out like this is to make clear that a very complex-looking neural network is still just a function with inputs and parameters, so we can treat it like every other model in this book. Equation (21.2) shows the network function of so-called *feedforward neural networks* with one hidden layer. The number of parameters in such networks is equal to

$$n_p = (n_x + 1)\, n_h + (n_h + 1)\, n_y,$$

where the first term gives the dimension of θ_h and the second that of θ_y. For other network types, such as *convolutional networks* (where neurons in the same layer can depend on different subsets of the preceding layer) and *recurrent neural networks* (which allow loops), we cannot write out the network function in such a simple way. Nevertheless, even those networks are functions, just like process-based models are functions that we represent algorithmically and not in the form of a formula.

Let us now examine the flexibility of neural networks: what range of system behaviours can they represent? We shall implement in R a similar example as was given by MacKay (1995) of a one-hidden-layer neural network with univariate input and output. We choose the hyperbolic tangent ($f_h(x) = \tanh(x)$) as the activation function for the neurons, and the identity function ($f_y(h) = h$) for the outputs. The hyperbolic tangent function is defined as $\tanh(x) = \frac{e^x - e^{-x}}{e^x + e^{-x}}$ so it maps input $x \in [-\infty, \infty]$ to $[-1, 1]$. It is a scaled version of the standard logistic $g(x) = \frac{1}{1+e^{-x}}$ (which maps to $[0, 1]$) such that $\tanh(x) = 2g(2x) - 1$. The size of our one-layer network is determined by the number of neurons in the single hidden layer. The network function depends on the weights and biases that we set for each neuron and for the single output. We can implement this specific network in R with very short code:

```
y <- sum( tanh(x*w1+b1) * w2 ) + b2
```

That single line of code is quite flexible, as it allows for any number of neurons in the hidden layer. The number of neurons is implicitly defined by the length of the vectors `w1`, `b1` and `w2`. Still, we prefer to have more general R-code where we can easily change the activation functions for the hidden layer and output. The following code does that, and it also has the nice property that it can accept a vector for `x` so that multiple input values are processed in parallel.

```
fNN.1n1 <- function( x, b1, b2, w1, w2, fh=tanh, fy=identity ) {
  X      <- cbind( 1, x )
  theta1 <- rbind( b1, w1 )     ; theta2 <- c( b2, w2 )
  h      <- fh( X %*% theta1 ) ; H      <- cbind( 1, h )
  y      <- fy( H %*% theta2 )
  return( y ) }
```

You see in the R-code that we choose the hyperbolic tangent as the default activation function for the neurons (`fh`) and the identity function for the output node (`fy`), but we give no default values for inputs `x`, biases `b1`, `b2` or weights `w1`, `w2`.

This type of neural network, with just one hidden layer, can represent any continuous function, if we have enough neurons. The top row of Fig. 21.3 shows

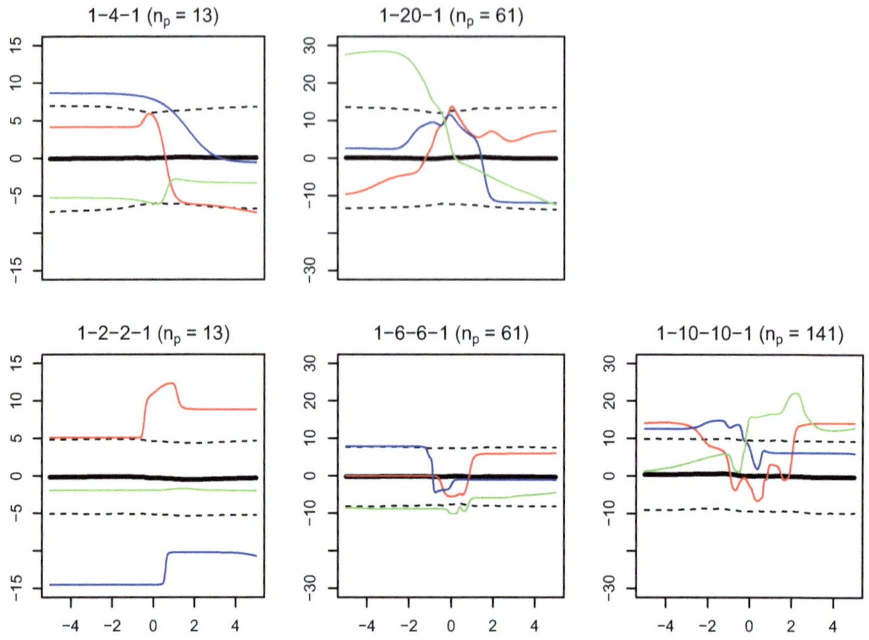

Fig. 21.3 Sampling from different neural network architectures, as specified in the headers, with the number of parameters between brackets. The number of hidden layers was one (top row) or two (bottom). Black lines: mean of 500 network functions sampled from the prior ± standard deviation. Priors were $N[\mu = 0, \sigma = 3]$ for all parameters. Coloured lines: three of the sampled network functions

examples of network functions generated by networks with a single hidden layer. The panel on the left is for an 1-4-1 network architecture, so it has four neurons in the hidden layer, whereas the one on the right has 20. We assigned the same zero-mean Gaussian prior to each parameter (weights and biases) with standard deviation equal to 3. From the joint prior, we generated 500 different network functions—the figure panels show three of them as well as the sample mean and standard deviation. Note the strong differences between the three sampled network functions for both architectures and the more complicated shapes generated by the 1-20-1 network.

21.2.2 Neural Networks with Multiple Hidden Layers

The number of parameters in fully connected feedforward networks with multiple hidden layers is equal to

$$n_p = \sum_{i=0}^{N} (n^{(i)} + 1)\, n^{(i+1)},$$

where N is the number of hidden layers and $n^{(0)} = n_x, n^{(1)} = n_{h1}, .., n^{(N+1)} = n_y$. (For $N = 1$ this reduces to the formula for one-hidden-layer neural networks mentioned in the preceding section.) The number of parameters scales roughly quadratically with the number of neurons, but it depends on the configuration. For example, 1-20-1 and 1-6-6-1 networks both have 61 parameters even though the first has over 50% more neurons.

Here is an R-function that implements feedforward neural networks with two hidden layers and univariate input and output:

```
fNN.1nn1 <- function( x, b1, b2, b3, w1, w2, w3, fh=tanh, fy=identity ) {
  n1     <- length( b1 ) ; n2 <- length( b2 )
  X      <- cbind( 1, x )
  theta1 <- rbind( b1, w1 )
  theta2 <- matrix( c( b2, w2 ), nrow=n1+1, byrow=T )
  theta3 <- c( b3, w3 )
  h1     <- fh( X  %*% theta1 ) ; H1 <- cbind( 1, h1 )
  h2     <- fh( H1 %*% theta2 ) ; H2 <- cbind( 1, h2 )
  y      <- fy( H2 %*% theta3 )
  return( y ) }
```

We used this R-function to study networks with two hidden layers. The bottom row of Fig. 21.3 shows the behaviour of three such networks with similar neuron numbers and priors as the one-hidden-layer networks that we studied before. You notice that the 1-10-10-1 network has the same number of neurons as the 1-20-1 network, but it generates more complex functions because it has more parameters.

We conclude that different network architectures can generate a wide variety of functions from their prior parameter distributions. This variety is a strength when we know very little about the real-world function that we aim to model, but can we constrain the functions enough to give us useful predictions? Let's see how well the networks can be trained using Bayesian calibration.

21.3 Regression Using a Neural Network

The first Bayesian calibration of a neural network using MCMC was reported by Neal (1992). He used Hamiltonian Monte Carlo on a 1-16-1 network. Here, we carry out Bayesian calibration of the same five neural networks whose prior predictive uncertainties we explored in Fig. 21.3. Our test data are $x = (-2, 0, 2)$, $y = (-10, 10, -1)$. We use the Metropolis algorithm for MCMC with the same R-implementation that we presented in Chap. 7.

The MCMC-generated sample from the posterior distribution for the 13 network parameters of the simplest one-hidden-layer network (1-4-1) is shown in Fig. 21.4. Figure 21.5 explores the posterior performance of all five networks. First note that the posterior mean parameter vector (red line) tends to perform poorly, as we would expect because the networks are nonlinear. The mean parameter vector only

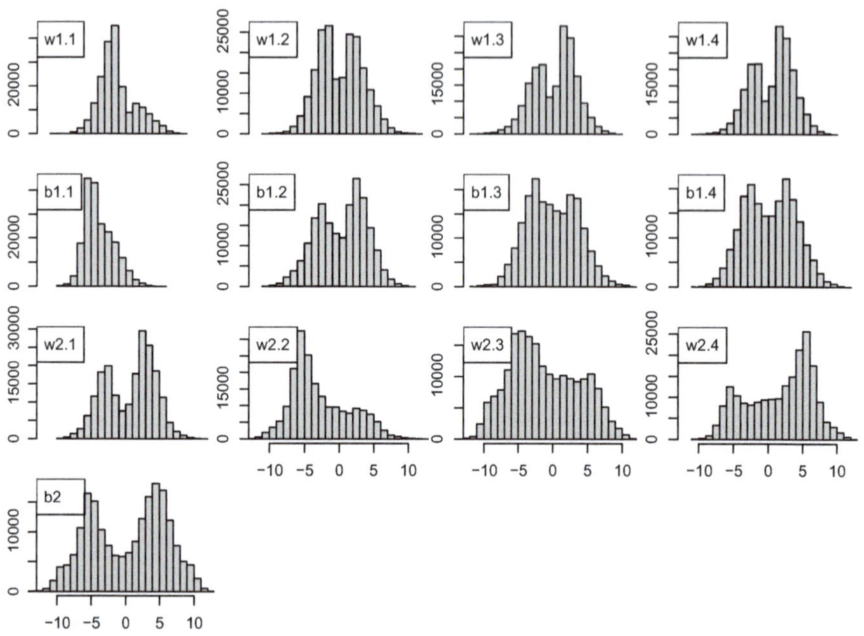

Fig. 21.4 Marginal posterior distributions for the 13 parameters of a 1-4-1 network

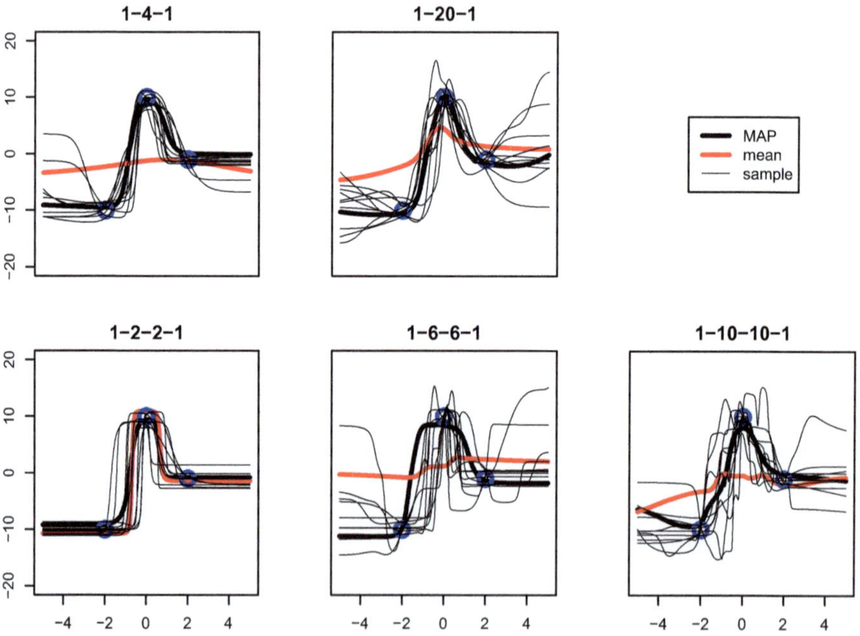

Fig. 21.5 Bayesian calibration of neural networks. Top row: one hidden layer. Bottom row: two hidden layers. Each line is generated by a different parameter vector: MAP (thick black), mean (thick red), random vectors from the MCMC (thin black)

generates a good fit in the case of 1-2-2-1 network, which is the most stable network anyway, as we see from the ten curves derived by subsampling the MCMC (thin black lines). But more representative than the mean is the parameter vector with highest posterior probability (the 'MAP' vector, thick black line) whose predictions are shown too. The MAP performs well in all cases. So the simple data that we used do not warrant a complex network. Moreover, the more complex networks allow for some wildly oscillating subsampled functions, which may be undesirable if we believe that the true relationship between x and y is not so wiggly. The use of neural networks, and much of machine learning in general, is a balancing act between desired flexibility and unwanted overfitting.

21.3.1 Preventing Overfitting

There are four obvious ways to constrain the behaviour of neural networks: (1) get more data, (2) restrict the values that the parameters can assume, (3) add a penalty term to the likelihood function that makes wiggly network functions unlikely, (4) limit the number of nodes and layers. All four ways have been tried with neural networks, but usually by means of ad-hoc methods (under headings such as 'regularisation' and 'penalisation') rather than Bayesian probability theory (James et al., 2013; MacKay, 2003).

Let's examine what the consequences in the above calibration example would be if we set the prior distribution for the neuron parameters (`w1`, `b1`) to be zero-centred as before, but reduce the variance by a factor 10. We try this narrow prior out with the 1-20-1 network which showed considerable wiggliness a posteriori as we saw in Fig. 21.5.

As Fig. 21.6 shows, with this low-variance prior the network produces mainly smooth functions, both a priori and a posteriori, with only minor reduction in its capacity to reproduce the training data. But remember that our dataset is atypically small. With richer datasets, we may need a wider prior and perhaps more neurons.

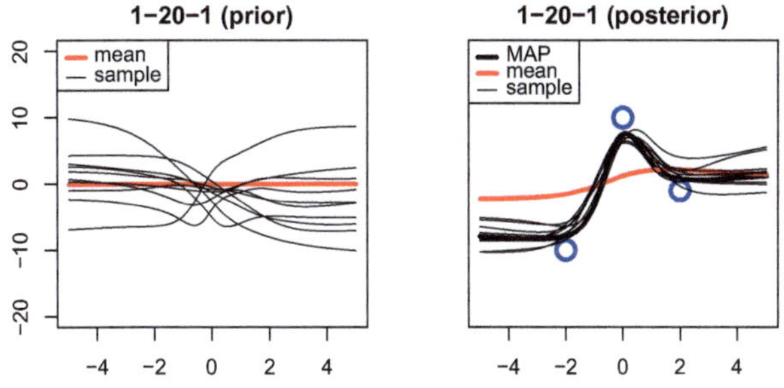

Fig. 21.6 Prior and posterior predictions of a 20-neuron one-hidden-layer neural network with narrow prior for weights and biases

However, a full Bayesian approach to derive (a representative sample from) the posterior distribution will then become difficult. It is possible to use approximation methods (Chap. 13), such as Laplace approximation (MacKay, 1992b) or variational Bayes (Bishop, 2006; Mann et al., 2022). But neural networkers usually do not even attempt that and restrict themselves to the estimation of the single parameter vector that is optimal for their training data, thereby sacrificing uncertainty quantification (Ghahramani, 2015). They use optimisation algorithms specifically designed for different types of network, such as the deterministic *backpropagation* gradient-descent algorithm for feedforward networks with multiple hidden layers (MacKay, 1992b; Rumelhart et al., 1986), or newer more widely applicable algorithms (Radhakrishnan et al., 2024).

21.4 Classification Using a Neural Network

Earlier, we used neural networks for regression of continuous output y on input x. But neural networks can also be used for classification. This usually takes the form of networks with multivariate output that can be interpreted as a discrete probability distribution. We encountered similar methodology in Chap. 20 in the form of generalised linear models (GLM) which likewise can be used for regression (e.g. when using the Gaussian distribution) and for classification (e.g. when using the binomial distribution as in logistic regression which, despite the name, is a classification tool). Neural networks can carry out both tasks because their individual neurons are GLMs. Indeed, we can view neural networks as 'extremely generalised linear models'.

The number of output nodes n_y in classifier networks is equal to the number of different classes. And to ensure that the outputs can represent a discrete probability distribution, their values are constrained to be non-negative and sum to 1, which is enforced by a suitable choice of activation function for the output layer. The value y_i of each output node is interpreted as the probability that the item belongs to the i-th class i.e. $p[z = z_i]$. Obviously, we like our network to assign values close to zero for all incorrect attributions, and a value close to 1 for the node associated with the correct class. So the likelihood function is the product of n_y Bernoulli probabilities:

NEURAL CLASSIFIERS likelihood function:

$$p[z|x, \theta] = \prod_{i=1}^{n_y} Br[z \equiv z_i | y_i], \tag{21.3}$$

where z is the true observed class of the item, y is the network's output for inputs x and parameters θ, and $Br[..]$ is each output node's Bernoulli probability of assigning the correct class label. $Br[z \equiv z_i|y_i]$ evaluates to y_i when the i-th output is associated with the true class z_i, and to $1 - y_i$ for all other outputs.

Let's now study an example of classification by neural network: deciphering handwriting.

21.4.1 Deciphering Hand-Written Digits

Deciphering human handwriting is a form of classification because the number of alphanumeric characters is finite. Our example here is about recognising hand-written digits. We use the famous MNIST dataset (Deng, 2012) that contains thousands of digital images of handwritten digits and can be downloaded from http://yann.lecun.com/exdb/mnist/ with the following R-command:

```
mnist <- dslabs::read_mnist()
```

Each image in MNIST is stored as a vector of 784 integers (grey values ranging from 0 to 255) representing a matrix of 28 × 28. We use two hundred MNIST matrices, half for training and half for testing. We block-average them such that each matrix is reduced to size 2 × 2. Figure 21.7 shows the first four training matrices, before and after block-averaging.

The block-averaging means that we are throwing away much relevant information and are left with very blurry pictures. But it allows us to use a smaller network, enough for demonstration purposes. It does make our recognition task much harder

Fig. 21.7 First four MNIST training matrices. Top row: original data (Deng, 2012). Bottom row: block-averaged. From left to right, the digits are 5, 0, 4, 1

of course! To facilitate parameter estimation, we scale the matrix elements to be in the interval [0,1].

Now we need to design our neural network. This is generally a matter of trial and error, within constraints. The network needs to have one input per matrix element, so four in total. And it needs to have one output per possible class (i.e. 0–9), so ten in total. We can choose any number of hidden layers and neurons, but we keep our network simple by having just one hidden layer of three neurons. So we are building a feedforward 4-3-10 network which has 55 parameters. As before, we use a `tanh()` activation function for the hidden layer neurons. For the output layer, we use the standard `softmax()` function ($e^{z_i} / \sum e^{z_j}$) which ensures that the 10 outputs are all in the unit interval and sum to 1, so they can represent a discrete probability distribution for the digit. In summary, for each input vector of length 4, the network generates an output vector of length 10 with the ten digit probabilities.

Now we prepare for calibration. As prior we choose Gaussians, with mean zero for the biases, 0.5 for the first-layer weights and 1 for the second-layer weights. Standard deviations are 3 for the output weights and 1 for the other parameters. The likelihood function is Eq. (21.3) with $n_y = 10$, and the MCMC is carried out with the same Metropolis algorithm that we used for neural network regression above.

After the calibration, we find that the network with the MAP parameter vector assigns the highest posterior probability to the true labels in 51% of training images. But the more important test is of course against the independent test images, and the percentage then drops to 40%. Still, this is a remarkably high value given that our images were obscured by the block averaging. Random assignment would have led to 10% correct identification. Figure 21.8 depicts the results graphically. It shows that blurry images of digit 1 are the easiest to identify and those of digit 5 the hardest.

But what about parameter uncertainty? When instead of only considering the MAP parameter vector, we take a representative sample from the posterior distribution for the network's parameters (by subsampling 20 parameter vectors from the chain generated by MCMC), we find that the percentage of correct digit identifications varies from 31 to 49%. So the MAP parameter vector did not provide exceptionally high predictive capacity, its performance is rather average. That is a common phenomenon with neural networks, which tend to have fairly flat likelihood functions in parameter space, so that quite different parameter vectors have similar performance. It makes them robust to parameter changes, but also harder to calibrate.

21.5 Sequence Extension and Natural Language Processing (NLP)

A very common application of machine learning is *sequence extension*. This is the problem of predicting the next item in a sequence z, when we know the last n items. If the sequence represents text in English or another human language, then we speak

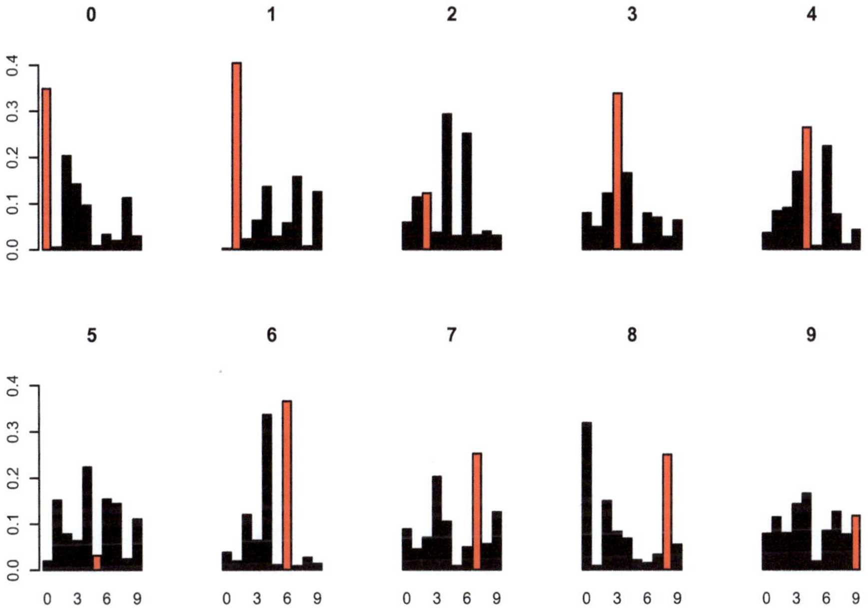

Fig. 21.8 Classification of MNIST digit-images using a 4-3-10 network with the MAP parameter vector from Bayesian calibration. Panel headers indicate true digit values. Bars represent the network's average probability of assigning the digits 0..9, in each case distinguishing correct (red) and incorrect attributions (black)

of *natural language processing* (NLP). In that case, the individual items are words, parts of words and interpunction, collectively called *tokens*. But sequence extension can also apply to computer languages and numerical sequences, the underlying ideas remain the same. The set of possible items is finite (let's call it the *vocabulary* as they do in NLP), so sequence extension is a problem of classification. We want to quantify the discrete probability distribution for the next item:

$$p[\, z(n+1) \mid z(1..n) \,],$$

where n is the *context length*. By iteratively sampling from this distribution, we can generate as many hypothetical sequence extensions as we want, and this is often referred to with the grand term *generative artificial intelligence (AI)*. A more modest name would be 'autoregressive machine learning'. The extension can be one-dimensional (text), two-dimensional (images), three-dimensional (videos), four-dimensional (videos with sound), but all are based on the same probabilistic principles. We shall restrict ourselves to one-dimensional examples in this chapter.

Let's begin with a numerical sequence to set the scene: the decimal expansion of $\pi = 3.14\ldots$ and say that our context length $n = 1$. Our classifier then is simply a *Markov Chain* defined by the transition probability matrix P where $P[i, j] =$

$p[z(n+1) = j \mid z(n) = i]$. The first n digits of the decimal expansion of π can be downloaded from the On-Line Encyclopedia of Integer Sequences as follows:

```
pi.url <- "http://oeis.org/A000796/b000796.txt"
pi.dec <- read.csv( pi.url, sep=" ")$X3[ 1:n ]
```

And here is part of the transition probability matrix P derived from the first 5000 decimals of π.

```
>      0    1    2    3    4    5    6    7    8    9
> 0 0.08 0.11 0.09 0.10 0.09 0.09 0.11 0.09 0.09 0.14
> 1 0.10 0.10 0.09 0.10 0.12 0.09 0.09 0.10 0.10 0.11
> 2 0.10 0.13 0.10 0.09 0.08 0.11 0.11 0.10 0.10 0.08
> ...
```

Every row of P sums to 1 and represents a multinomial distribution over the 10 possible next decimals. We derived each row's 10 probabilities by normalising the frequency distribution of digit transitions in our 5000-digit long training set. But what is Bayesian about that procedure? Well, the approach was similar to our Bayesian calibration of hazard probability that we described in Chap. 18. There we used a Beta prior and its conjugate binomial likelihood function to arrive at an analytical expression for the Beta posterior. Here the parameters θ that we want to estimate are also probabilities, but we now have 10 of them for each row, rather than just one. So we may want to use a Dirichlet prior $\mathrm{Di}[\theta|\alpha]$ and a multinomial likelihood $\mathrm{Mu}[\theta|z, n]$, which are multidimensional generalisations of the Beta and binomial. These distributions are also conjugate so there is an analytical expression for the Dirichlet posterior $\mathrm{Di}[\theta|\alpha']$ (see e.g. Bernardo & Smith, 1994). We can choose the parameters $\alpha = \{\alpha_i\}$ of the Dirichlet prior to all be equal and close to zero, representing great uncertainty and ensuring that the posterior is dominated by the training data and resembles the normalised observed frequencies. The elements of our transition probability matrix are the expectation values under the resulting posterior distribution. To keep our exposition short, we'll now focus on those expectation values and not consider the uncertainty of the transition matrix.

All 100 probabilities of the full transition probability matrix that we found in this way are in the range [0.08,0.14]. Yet the number π itself is of course not random, and there are many deterministic algorithms that calculate its decimal expansion. But π is an irrational number whose expansion is a 'normal sequence' in the technical sense that digit subsequences of any length are uniformly distributed. Because of this, all elements of our transition matrix are close to 0.1, and with a larger sample size they would have been even closer. Using our matrix to predict decimals 5001, 5002, etc. of π would therefore lead to a nearly uniformly random sequence extension. In fact, no current-day machine learning algorithm can find useful patterns here. We posit the 'incompleteness theorem of machine learning': there exist deterministic patterns in nature that can never be generated by any machine learning algorithm!

Having said that, machine learning has proved far more powerful than was expected only a few years ago. A prime example is its application to sequence extension in NLP. So let's explore auto-completion of text. We shall train our algorithm on Bertrand Russell's 'The Problems Of Philosophy' (Russell, 1912). This is one of many books that can be freely downloaded into R from the Gutenberg repository:

```
f   <- "https://www.gutenberg.org/cache/epub/5827/pg5827.txt"
txt <- scan( f, "char", skip=66, blank.lines.skip=T, nlines=1000 )
```

That R-code reads the first 1000 lines (11,055 words) of Russell's book, skipping 66 lines of introductory material. That will be our training text, and here are the first two sentences:

> *Is there any knowledge in the world which is so certain that no reasonable man could doubt it? This question, which at first sight might not seem difficult, is really one of the most difficult that can be asked.*

Good to see that Russell also stresses the importance of thinking about uncertainty! Our task will be to find a plausible continuation in Russell's style. We make our life a little easier by some pre-processing of the text. We remove all special symbols and most interpunction, leaving only whole words and the full stop. Despite the inclusion of the full stop, we shall refer to all classes (tokens) as 'words'. The word vocabulary that is thus found is shown in Fig. 21.9.

For the text sequence extension, we use the same Bayesian Markov Chain approach that we employed, without success, in the expansion of π. Andrey Markov himself was the first to use this approach to text generation in 1913 when he analysed Pushkin's narrative poem 'Eugene Onegin' (Hayes, 2013)! His tokens were letters rather than words. But back to Russell—this time we do not make the transition

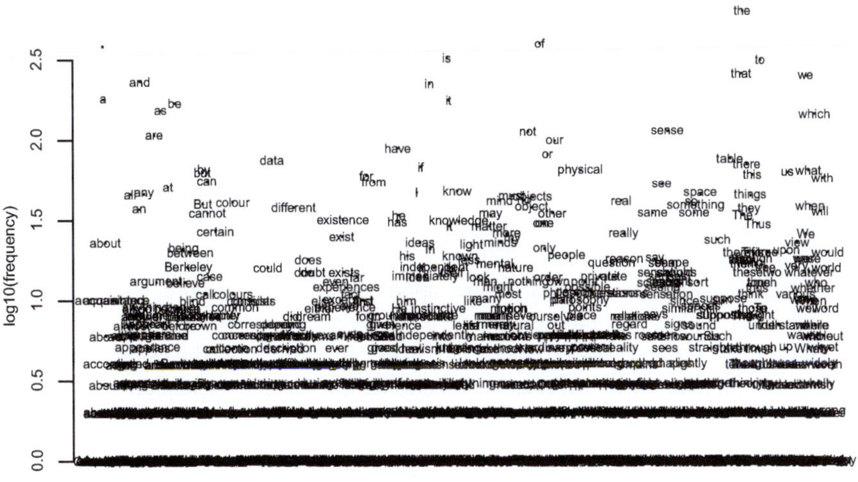

Fig. 21.9 Vocabulary of the first 1000 lines of Russell (1912). The 1552 different words are sorted alphabetically from left to right. Vertical position is log10(word frequency)

probability matrix explicit (it is too big at 1552 × 1552) but calculate the transition probabilities on the fly, by identifying the successors to the last word in the training text, and their frequencies. The last word thus implies a multinomial distribution for its own successor. We prompt our text extension with a full stop, sample a new word from the implied distribution and keep repeating that process to extend our text. Here is the new text that is generated that way:

> *Hence, we shall be acquainted with other sense data were any rate no difficulty in mind and also held among philosophers are to perceive the history nobody comes . Leibniz says that the relations of sight of being known except such things that there must be something existing independently of sense data but afterwards cat consists of detail Berkeley. On necessarily the only upon doubtful some sense data is a thought of space we any given moment no hunger becomes when I can be supposed them . It seems to the distinction the same applies to draw that*

Despite the absurdly small context length, this generated text is not complete gibberish, more like a half-overheard conversation. But for better results, we obviously need greater contexts. And that brings us to the question, how are modern NLP algorithms implemented? Well, in most cases, complex neural networks are used rather than a minimal Markov Chain. Despite that difference, the general idea remains the same: train your algorithm on data to produce a discrete probability distribution for the next word (or token) and use that distribution to extend your text. Let's briefly discuss the developments in such *large language models* (LLM).

The vocabulary size of LLM is in the order of 10^5 (Naveed et al., 2024). So a simple transition probability matrix with a context length of 1, like the one we constructed above, would already contain $\sim 10^{10}$ parameters. Above we circumvented any problems with this by carrying out the training on the fly, word by word. That was feasible because our training text (the first 1000 lines of Russell, 1912) could easily be kept in computer memory and consulted at every step, but this is not feasible if a LLM is trained on large parts of the internet. There are attempts to link chatbots (LLMs that can be prompted by user questions) to advanced search engines (e.g. Wolfram, 2023), and this will likely become a standard procedure. The context length of LLMs is generally around 2^{12} but some are already reaching 2^{21} (Li et al., 2024). To use contexts efficiently, some parts of them, such as the user prompt and specific key words, tend to receive extra attention (Bengio et al., 2003; Vaswani et al., 2023). But in the world of machine learning, no solution is permanent. Novel network architectures for processing large vocabularies and contexts keep being introduced (Zhao et al., 2023). The number of parameters of LLMs keeps increasing and is now approaching 10^{12} (Zhao et al., 2023). So the computational and time demand of parameter estimation remains a concern. Therefore, in general, no attempts are made to carry out Bayesian parameter calibration on neural networks, as we mentioned earlier in this chapter.

21.6 Classification Using Naive Bayes (NB)

Naive Bayes is a classification method that assumes that all data are independent. So the likelihood function is the simple product of the likelihoods for the individual pieces of information (referred to as 'features') about the to-be-classified item.

NAIVE BAYES likelihood function:

$$p[y|z_j, \theta] = \prod_{i=1}^{n_y} p[y_i|z_j, \theta_{i,j}], \tag{21.4}$$

where $y = \{y_i\}$, $i = 1..n_y$ are the observed features of the item, and its class is denoted by $z_j \in 1..n_z$. Individual model parameters $\{\theta_{i,j}\}$ are feature- and class-specific, so the properties of the probability distribution $p[y_i|z_j, \theta_{i,j}]$ depend on the feature in question and the given class. For binary features, we can use a Bernoulli distribution, for categorical features a multinomial, and for real-valued features a Gaussian. The latter application is referred to as the Gaussian Naive Bayes (GNB) classifier, and we have already encountered its likelihood function in Eq. (4.3), in the context of regression rather than classification. The Bernoulli and multinomial NB classifiers are often used in natural language processing, e.g. in the detection of SPAM emails.

Let's see Naive Bayes in action. We use Anderson's famous dataset with 150 records on iris flowers (Anderson, 1936) that is often used to demonstrate data analysis algorithms. It is a built-in dataset in R. There are 3 classes (iris species, stored in vector z) and 4 features (leaf sizes, stored in matrix y). We use half the records for training, i.e. estimating species-specific means and variances for the features. The other half will be used to test the trained classification. Here is our R-code that prepares all that:

```
nz      <- 3                       ; ny     <- 4
z       <- as.numeric(iris[,5])    ; y      <- as.matrix(iris[,1:ny])
r.train <- c(1:25,51:75,101:125)   ; r.test <- setdiff(1:150,r.train)
y.train <- y[ r.train, ]           ; y.test <- y[ r.test, ]
z.train <- z[ r.train  ]           ; z.test <- z[ r.test  ]

m.train <- s.train <- vector("list",nz)
for( j in 1:nz ) {
  m.train[[j]] <- apply( y.train[z.train==j,], 2, mean )
  s.train[[j]] <- apply( y.train[z.train==j,], 2, sd   ) }
```

We assign a uniform prior for the species, so the Bayesian posterior for class z amounts to relative likelihoods. We use Gaussian Naive Bayes here because of the real-valued leaf sizes and implement the naive likelihood as an R-function that we call `L.iris`:

```
L.iris <- function( y, z, m=m.train, s=s.train ) {
  prod( dnorm( y, m[[z]], s[[z]] ) ) }
```

The likelihood function arguments are the observed features `y`, a given class `z`, and the lists `m` and `s` that for each of the three classes contain the four feature means resp. standard deviations estimated from the training data. You see from the second line of the code that this likelihood function simply multiplies the four feature-specific univariate Gaussian probability densities.

When we apply this function to the test data, and assign each flower to its most probable species, we get the results shown in Fig. 21.10, left panel. The Naive Bayes algorithm performs very well for this dataset, with only three errors, so the accuracy is 72/75. The three flowers it got wrong were amongst those for which it was not very certain anyway ($p[z] < 0.85$).

Because of the high accuracy of Naive Bayes in this case, there is not too much point in examining the results more closely. But in general, you can summarise the performance of a classification in a *confusion matrix*. This is a matrix C of size $n_z \times n_z$, where C_{ij} is the fraction of true class i items put in class j by the algorithm. So ideally the values on the diagonal are close to 1. A confusion matrix can be a convenient tool to analyse problems with the classification. You may for example want to know which class z_i we get wrong the most. And is it by falsely classifying an item as z_i, or by falsely classifying a true z_i as something else? In our iris example the three errors that Naive Bayes made were all confusing species 2 and 3.

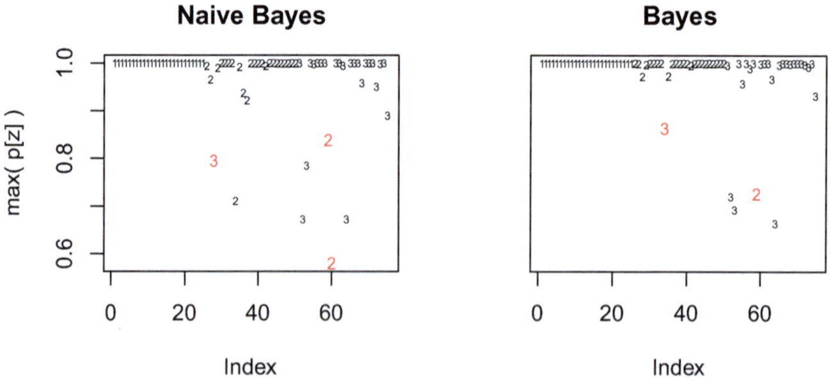

Fig. 21.10 Gaussian classification of iris flowers. Left: covariances ignored. Right: covariances accounted for. Digits 1–3 indicate the most probable species for each of 75 tested plants. Classification is correct (black digits) or incorrect (large red digits)

We now ask, would a less naive Bayesian classification have classified the irises better than Naive Bayes? Let's repeat our classification exercise with a likelihood function that does acknowledge correlations between features. So we replace, for each species, the four feature-specific standard deviations with a single covariance matrix that we estimate using the training data. So our new likelihood R-function (`L.iris.S`) calculates a single multivariate Gaussian probability density for each flower:

```
L.iris.S <- function( y, z, m=m.train, S=S.train ) {
  dmvnorm( y, m[[z]], S[[z]] ) }
```

With this more fully Bayesian approach, which does not ignore relevant information on correlations, we get the classification results shown in the right panel of Fig. 21.10. The accuracy has increased to 73/75. But don't read too much into this slight improvement. The accuracy of classification algorithms depends not only on the algorithm itself but also on the variation between classes and the quality of the training data. Naive Bayes has often been found to be surprisingly effective, but whenever possible do choose the full Bayesian approach because it uses all the information you have and thus provides more reliable estimates and uncertainty quantification.

21.7 Outlook for Machine Learning

Our main example of neural networks showed how machine learning starts from highly flexible function definitions which it aims to constrain using data. We could have chosen other machine learning formalisms, such as random forest modelling, to illustrate the flexibility as well as the associated problems of parameterisation. In machine learning, these problems have often been swept under the carpet by not attempting to derive a full posterior probability distribution for the parameters, but instead just identifying a single well-performing parameter vector through minimisation of a *cost function* (pseudo-likelihood). Parameter uncertainty is then not assessed, but predictive uncertainty can still be approximately quantified using cross-validation (splitting the data into training and validation data in multiple configurations). Despite the continuing prevalence of these ad-hoc methods for parameterising machine learning models, Bayesian methods are slowly being introduced. There are, for example, R-packages for Bayesian regression tree modelling (Kapelner & Bleich, 2016) and in 'caret', a powerful general-purpose R-package for machine learning, various Bayesian methods are included (Kuhn, 2008).

Are there no other functions that are as flexible as the typical machine learning functions yet facilitate uncertainty quantification? Well, at various places in this book (e.g. Chap. 15) we have encountered *Gaussian Processes* (GP) which seem to fit the bill. In fact, Neal (1996) showed that the prior predictive uncertainty represented by the output from an untrained feedforward neural network with

Gaussian priors for the weights and biases (as was used to produce Fig. 21.3) will—in the limit of an infinite number of neurons—be equal to the predictive uncertainty of a GP. Rasmussen and Williams (2006) showed in detail, and from a Bayesian perspective, how GP can be used for machine learning tasks, including regression and classification.

Another application of GPs in machine learning is in so-called *Bayesian optimisation* of the 'hyperparameters' of complex models. These are not the hyperparameters that define prior probability distributions in Bayesian hierarchical modelling (Chap. 17), but parameters that control the training of machine learning models, such as the number of decision trees in random forests and the learning rate in neural networks. So this is yet another example of terminological confusion! Bayesian optimisation by GP consists of emulating the likelihood function of a complex machine learning model over its hyperparameter space. The emulator is gradually improved in an iterative procedure whereby at each step the hyper-parameter values that are optimal according to the emulator are supplied to the complex machine learning model and tested against data. That new complex model evaluation then helps improve the emulator. We discussed emulation of complex functions by means of GP at length in Chap. 15. Gaussian process calculations require the computationally demanding procedure of matrix inversion, so they may not always be a good substitute. Still, Bayesian machine learning using GP remains a subject of active research. For an overview of much recent work, see Murphy (2023).

At the beginning of this chapter, we mentioned another problem with machine learning: it is a black-box approach that leads to algorithms that are difficult to interpret. To ameliorate the problem, there are efforts to insert causal modelling into some of the layers of deep learning networks (Bradley et al., 2022; Reichstein et al., 2019). Essentially, this is hybridising machine learning with process-based modelling. We generally use process-based modelling when we understand a system's mechanisms well, whereas machine learning is used when we have many data but little understanding. Hybridising the approaches is a highly intriguing idea that will hopefully lead to algorithms that are not only easier for humans to understand but also have greater predictive capacity in conditions that differ from those prevailing during collection of training data. As Rudin et al. (2022) advocate, building interpretable machine learning models in the first place may be preferable to trying to explain black-box models a posteriori.

We can also imagine going beyond hybridisation and replacing the process-based models partly or completely by machine-learned algorithms. If the substitute models are trained on the input–output relationships of the PBMs, then this is called model emulation. Climate models, which are notoriously slow to run, have been efficiently emulated by machine-learned models with speed increases of two to six orders of magnitude (Wong, 2024). We briefly discussed data-driven vs. mechanistic model emulation in Chap. 15. Lam et al. (2023) showed remarkable improvements in world-wide 10-day weather forecasting by training a neural network on output from a process-based global weather forecasting model. Their approach does not yet include uncertainty quantification, but they see that as a 'crucial next step'.

Bayesian weather forecasting is a topic of research, not yet of practical application (Liu et al., 2020). A more modest goal than replacing a PBM altogether would be to only represent model structural error ('discrepancy') by means of a machine-learned algorithm (see Chap. 12).

In classification, as in all modelling, the Bayesian approach entails specifying a prior distribution for the parameters and a likelihood function for the observations, deriving the posterior parameter distribution, and acknowledging posterior predictive uncertainty. So for each item that we want to classify based on its features, we have a posterior distribution for the possible classes. But if we have to simplify this stochastic model to a deterministic classification rule, then we would obviously say 'assign the item to that class for which, given the item's features, the posterior probability is the highest'. That deterministic rule will on average make the fewest errors and is called the *Bayes optimal classifier*, or Bayes classifier for short. The terminology is unfortunate because the truly Bayesian thinking takes place before the deterministic simplification.

Finally, a brief note on the application domain of machine learning. In the context of the calculation of the decimal expansion of π, we posited the *incompleteness theorem of machine learning* stating that there are deterministic patterns in nature that can never be generated by any machine learning algorithm. We may know that the pattern is deterministic, and we may even have a mathematical algorithm for it, but machine learning will never be able to hit on the algorithm by training on data alone. Even if this theorem is true, the application domain of machine learning will continue to expand. In the case of π, we can circumvent it by inserting more mathematical knowledge into machine learning, e.g. by the hybrid approach mentioned above. And for example in geometry, certain types of theorems can already be discovered and proven by neural networks (Trinh et al., 2024). Machine learning will become an indispensable tool for the majority of mathematicians and scientists. Its main weakness will remain the difficulty to apply Bayesian methods to parameter-rich machine learning algorithms, thereby hampering uncertainty quantification.

Exercises

1. **Classifying disease.** Consider the medical example of Chap. 2. Suppose that after your positive test result you decide to have another test taken, so you now have two test results. Do you think that Naive Bayes would be a good tool to classify yourself as either infected or not?
2. **The neural network of Fig. 21.2.** How many parameters does this neural network have? Use the formula provided in this chapter.
3. **1-1-1 neural networks.** The simplest neural networks are 1-1-1 networks. How could you specify in R a 1-1-1 network that just doubles its input? Use the R-function `fNN.1n1` that we defined in the chapter.

Chapter 22
Time Series and Data Assimilation

Time series modelling can be defined as modelling the progression over time of the state of a dynamical system. We focus here on simple statistical models rather than complex process-based models (PBMs) but will come back to those at the end of this chapter (and for more about PBMs see e.g. Chaps. 8, 11 and 12).

From a Bayesian perspective, we always express what we want to predict as a conditional probability distribution. The future is conditional on the present, and our uncertainties about the present will affect the quality of our prediction. So all time-series modelling is a matter of specifying the probability distribution $p[future|present]$ or $p[z(t+1)|z(t)]$. And in common time series practice, uncertainties tend to be expressed as Gaussian distributions. That leads to the question: can we use Gaussian Processes, which we encountered a few times already, to represent common time series models? Let's begin there.

22.1 Sampling from a Gaussian Process (GP)

We generally use a GP to represent our uncertainty about the shape of a continuous function f (Chap. 15). The domain A of f can be unbounded and multidimensional ($A = \mathbb{R}^n$) or just a line segment ($A \subset \mathbb{R}$). In all cases, A is an infinite set of points over which the GP defines a joint Gaussian probability distribution for $f(s), s \in A$. The GP is fully defined by its mean and covariance functions $m(s)$ and $C(s, s')$. But how exactly does the GP represent our uncertainty about $f(s)$? How can we produce a representative sample of possible function shapes? We can do this step for step by first sampling the GP at a single point, then at another conditional on the value we sampled for the first point, and so on. That iterative sampling procedure uses the decomposition that we can write for every joint distribution: $p[F] = p[f_0]\, p[f_1|f_0] \ldots p[f_n|f_0, \ldots, f_{n-1}]$. We can of course only sample at a finite number of points, so this does not fully determine the shape of the function.

M. van Oijen, *Bayesian Compendium*,
https://doi.org/10.1007/978-3-031-66085-6_22

How many points we need for a good approximation depends on the expected wiggliness of the function which is determined by the covariance function $C(s, s')$. Let's give an example for a function f whose one-dimensional domain A is the interval from 0 to 30. We represent our uncertainty about f as a GP with mean function $m(s) = 0$ and covariance function $C(s, s') = 3\exp(-|s - s'|/10)$. This choice of GP means that we expect $f(s)$ to be centred around zero with an overall variance of 3 and correlations between points that decrease exponentially with distance at a correlation length of 10. This is a much simpler GP than we introduced in the chapter on model emulation, but we could still use the general R-function for GP-prediction that we implemented in that chapter and which we named GP.pred. But let's write, for present use, a simplified version of that R-function so you do not need to look it up (although you might find it a good exercise to compare the two versions):

```
GP.pred.0 <- function(x0,x,y,Sy,phi) {
  C0    <- Sy[1] * exp( -abs(x-x0)/phi )
  m0_y <- t(C0) %*% solve(Sy) %*% y
  S0_y <- Sy[1] - t(C0) %*% solve(Sy) %*% C0
  return( list( "m0_y"=m0_y, "S0_y"=S0_y ) ) }
```

Now we are ready to start sampling from our GP. We assume that the domain of the function, which is the continuous interval [0,30], will be adequately covered if we sample at $x = 0, 1, \ldots, 30$. We start at $x_0 = 0$, where we generate a value for $y_0 = f(x_0)$ from the Gaussian $N[m(0), C(0, 0)] = N[0, 3]$ and then we keep applying our R-function GP.pred.0 to sample each following point. When we have reached x_{30} and produced a value for y_{30} we will have produced one sample 'function'. To sample representatively from our GP, we do this 1000 times. The following R-code does the sampling.

```
set.seed(13)
Vy   <- 3 ; phi <- 1e1 ; nseries <- 1e3 ; n <- 30
Y        <- NULL
for(i in 1:nseries) {
  x    <- 0 ; y   <- rnorm(1,0,sqrt(Vy))
  for(xi in 1:n) {
    dx <- as.matrix( dist(x) ) ; Sy <- exp( -dx/phi ) * Vy
    py <- GP.pred.0(xi,x,y,Sy,phi)
    yi <- rnorm(1, mean=py$m0_y, sd=sqrt(py$S0_y))
    x  <- c(x,xi) ; y <- c(y,yi)
  } ; Y <- cbind( Y, y )
} ; mY <- rowMeans(Y) ; sY <- apply(Y, 1, sd)
```

The results of the GP-sampling are summarised in the left panel of Fig. 22.1. As expected, our 1000 functions are mostly close to zero, with a standard deviation of $\sqrt{3}$. The individual functions in our sample, of which three are shown, differ a lot from each other, but all are quite wiggly.

You may well ask, what has all that to do with time series? Well, if we reflect on how we generated each sampled function, our procedure mimicked a temporal process. We kept moving our function evaluation from one point on the x-axis to the next. And that is what statistical time series are: they are rules for generating the value of a function $z(t)$ as a function of the value the function had in the past, e.g. $z(t-1)$. Specifically, the time series mimicked by our GP is called an *autoregressive* time series with dependency on the last value, denoted as an *AR(1)* process. The general formula for an AR(1) process is:

$$\underline{\text{AR(1) time series model:}}$$
$$z(t) = \alpha\, z(t-1) + \epsilon(t) \quad \text{where} \quad \epsilon(t) \sim N[0, \sigma_\epsilon^2]. \tag{22.1}$$

If $|\alpha| < 1$, the AR(1) model is stationary such that our uncertainty about any future value of z is independent of t and captured by a Gaussian distribution: $z(t) \sim N[0, \sigma_\epsilon^2/(1-\alpha^2)]$. And the AR(1) model then has its own covariance function $C_{AR(1)}(t, t-k) = \alpha^k \sigma_\epsilon^2/(1-\alpha^2)$ (from which of course the variance follows by plugging in $k = 0$).

We now see that the AR(1) is equal to our previous GP if both use the same covariance function. We achieve that by setting $\alpha = \exp(-1/\phi)$ and $\sigma_\epsilon^2 = 3(1-\alpha^2)$. We can then replace the GP-code with the simpler AR(1)-code without changing the underlying statistical model. Here is the code for sampling 1000 time series from the AR(1) process that should be equivalent to our GP:

```
set.seed(13)
alpha <- exp(-1/phi) ; Ve <- Vy * (1-alpha^2)
Z <- NULL
for(i in 1:nseries) {
  z <- rnorm( 1, 0, sqrt(Ve/(1-alpha^2)) )
  for(t in 1:n){
    z1 <- rnorm(1, mean=alpha*tail(z,1), sd=sqrt(Ve))
    z  <- c(z,z1)
  } ; Z <- cbind( Z, z )
} ; mZ <- rowMeans(Z) ; sZ <- apply(Z, 1, sd)
```

This is much faster code because it does not involve inversion of an ever-expanding covariance matrix for already examined points, as we have in the GP sampling. A summary of the 1000 time series produced by our AR(1) model is shown in the right panel of Fig. 22.1. The results are in every detail identical to the GP-results on the left! This is because the underlying mathematics is equivalent and the sampling for both models, which involved 31,000 calls to R's random-number generator (as you can verify from the code), was initialised with the same seed.

The AR(1) model can easily be made more general. We can add a non-zero intercept c and extend the memory of the system further back than the preceding step. That gives us AR(p) models, defined as $z(t) = c + \sum_{i=1}^{p} \alpha_i\, z(t-i) + \epsilon(t)$. We can also introduce a memory of the noise term, which leads to *moving average*

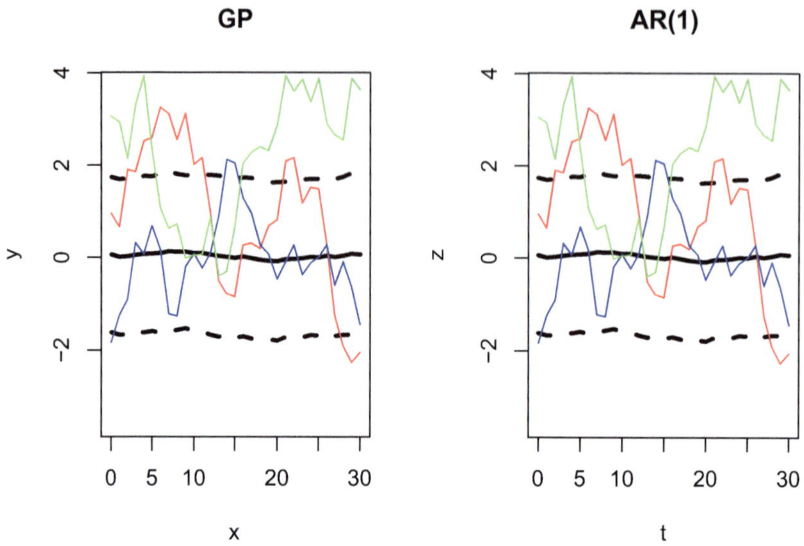

Fig. 22.1 Left: sampling from a Gaussian Process. Black lines are mean plus-minus one standard deviation of 1000 sampled functions for each value of x. Coloured lines show three sampled functions. Right: as on the left but from an AR(1) process rather than a GP. The same random seed was used for both panels

time series models, denoted as MA(q). Combining the two model types gives us the quite flexible ARMA(p,q) models, defined as:

ARMA(p,q) time series model:

$$z(t) = c + \sum_{i=1}^{p} \alpha_i \, z(t-i) + \sum_{j=1}^{q} \beta_j \, \epsilon(t-j) + \epsilon(t) \quad \text{where} \quad \epsilon(t) \sim N[0, \sigma_\epsilon^2]. \tag{22.2}$$

All these can be seen as examples of *state-space models* that are special cases of GP (Williams & Rasmussen, 1996). This is just a reminder of the fact that a huge part of statistical modelling consists of combining linear relationships with Gaussian probability distributions in different ways. And it means that time series modelling does not add new ideas beyond what was already implicit in the GP that we have found useful in various applications. It all boils down to different ways of specifying and using the covariance function for a multivariate Gaussian distribution. However, time series modelling can be a very efficient choice, as we saw in the much higher speed of the AR(1) code compared to its equivalent GP counterpart. And the efficiency of time series modelling is maintained when we extend it with assimilating information from observations at each time step, as we investigate in the following section. That will also bring us back to truly Bayesian methods.

22.2 Data Assimilation Using the Kalman Filter (KF)

Data assimilation is the name commonly given to methods that keep alternating between predicting the future state of a system and correcting the state estimate using new data. So data assimilation uses data to learn about the *state* of a system, whereas Bayesian calibration uses data to learn about the *parameters* of a model. The distinction is not very important as the underlying mathematics is the same, both involve modelling, and data assimilation algorithms tend to be Bayesian too, as we shall see.

Let's introduce some notation. In data assimilation, the future state z of a system is predicted using a model f, followed by using data y to correct that estimate. If the f in question is a state-space model, then data assimilation is a natural extension to the time series models that we have discussed in this chapter. Another term for data assimimilation of time series models is *filtering*. And if we stay in the Gaussian world of linear state-space modelling then the classical example of data assimilation is *Kalman filtering* (KF).

The KF in its richest formulation is a versatile data assimilation tool, but we begin by presenting a simplified KF that can only be used when the system z is univariate (i.e. at any time t, the system state $z(t)$ is a scalar, not a vector). We also assume for now that the state is directly observable, albeit noisily: $y(t) \sim N[z(t), \sigma_y^2]$. Finally, we assume that each application of model f multiplies the last state estimate by a constant α and adds Gaussian uncertainty with constant variance σ_f^2. The KF is then defined as follows:

KALMAN FILTERING (simple case of observable scalar state):

$$
\begin{aligned}
&\text{'Kalman Gain': } K = \sigma_p^2/(\sigma_p^2 + \sigma_y^2).\\
&\text{Data assimilation step: } p[z_a|\mu_p, \sigma_p^2, y, \sigma_y^2] = N[\mu_a, \sigma_a^2],\\
&\qquad \text{where } \mu_a = \mu_p + K(y - z_p) \text{ and } \sigma_a^2 = (1 - K)\sigma_p^2.\\
&\text{Prediction step: } p[z_p|\mu_a, \sigma_a^2] = N[\mu_p, \sigma_p^2],\\
&\qquad \text{where } \mu_p = \alpha\, \mu_a \text{ and } \sigma_p^2 = \alpha^2\sigma_a^2 + \sigma_f^2.
\end{aligned}
\tag{22.3}
$$

In Eq. (22.3), I have left out the time indices to avoid cluttering, but don't forget to read the equations as steps that we keep repeating as we move through time. The following R-code implements this simple-case KF-algorithm, with α set to 0.9, the model variance σ_f^2 set to 1, and some arbitrary data added.

```
set.seed(13)
n     <- 30 ; y <- runif(n,0.8,1) ; Vy <- 3
zp    <- Vp <- za <- Va <- rep(NA,n)
zp[1] <- 0 ; Vp[1] <- Vf <- 1 ; alpha <- 0.9
for(i in 1:n) {
  Ki       <- Vp[i] / ( Vp[i] + Vy )
  za[i]    <- zp[i] + Ki * ( y[i] - zp[i] )
  Va[i]    <- ( 1 - Ki ) * Vp[i] ; if(i==n) break
  zp[i+1] <- alpha   * za[i]
  Vp[i+1] <- alpha^2 * Va[i] + Vf }
```

The results of this simple KF are shown in Fig. 22.2. The figure demonstrates some quite general properties of data assimilation. We see that after each model prediction of the system state ($z_p \sim N[\mu_p, \sigma_p^2]$), the assimilation of a new data point acts as a correction on that estimate, so that the $\mu_a = E[z_a]$ lie between model estimate μ_p and observation. We also see that our uncertainty about the state estimate is lowest after a data assimilation ($\forall t : \sigma_a^2(t) < \sigma_p^2(t)$). That is because the predictions add model uncertainty. If observational and model uncertainties are constant as here (σ_y^2 and σ_f^2 are independent of t and z), then the variance of our state estimate will gradually start oscillating between constant values for σ_a^2 and σ_p^2. This happens because the repeated increase from σ_a^2 to σ_p^2 (= $(\alpha^2 - 1)\sigma_a^2[i] + \sigma_f^2$)

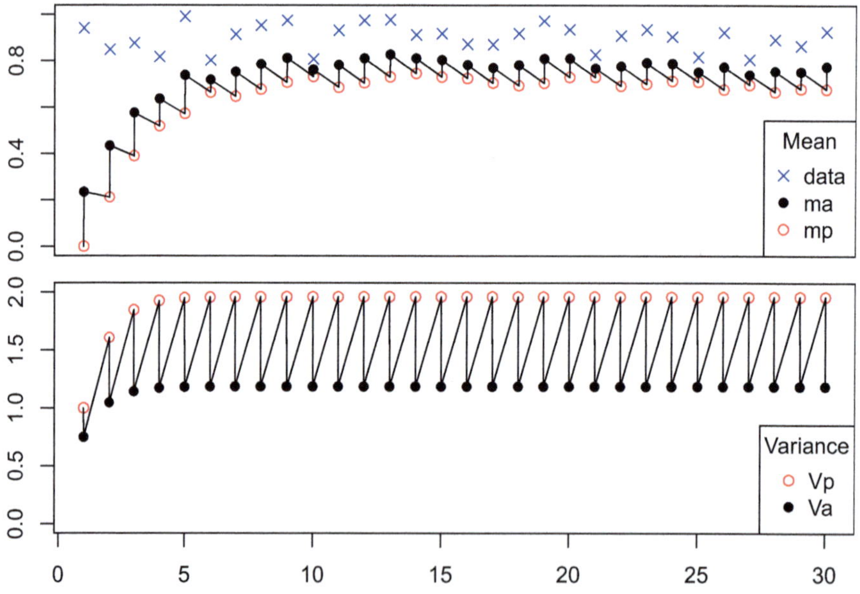

Fig. 22.2 Data assimilation using the KF. The top panel shows the observations and the alternating state estimates from model prediction (mp) and data assimilation (ma). The bottom panel shows the corresponding uncertainties as variances. Observational uncertainty is not shown but has a constant variance equal to 3

becomes equal to the decrease from σ_p^2 to σ_a^2 (= $K_i\,\sigma_p^2[i]$). The fact that variable data keep coming in does not stop the variances from stabilising because—as you can see from Eq. (22.3)—the (changes in) variances do not depend on the *values* of the observations, only on their *uncertainty* which we assumed to be constant.

So how would the KF behave if our data were not very reliable? Well, we can check what happens when the data are not informative without changing our R-code by setting the data variance σ_y^2 to a very large value, say $\sigma_y^2 = 10^8$. If we do that, and also set $\alpha = \exp(-1/10)$ and $\sigma_f^2 = 3(1-\alpha^2)$, then we are back to our GP & AR(1) example of earlier in this chapter, and the variances then become the same as we found there (Fig. 22.1): $\forall t : \sigma_p^2(t) = \sigma_a^2(t) = 3$. This confirms that KF is just time series modelling extended with data assimilation at each time step.

22.2.1 A More General Formulation of KF

We now look at a more complete formulation of the KF: see Eq. (22.4). The first difference with our earlier, simplified implementation of the KF (Eq. (22.3)) is that all quantities are now vectors and matrices rather than scalars. So the state of the system $z(t)$ and the observations $y(t)$ can be multivariate, in which case uncertainty about their values is expressed as covariance matrices (Σ_a, Σ_p, Σ_y) rather than scalar variances (σ_a^2, σ_p^2, σ_y^2). Similarly, the model prediction step has changed from multiplying the state variate with a constant scalar α to multiplication with a model matrix $\mathbf{F}$, with associated predictive uncertainty matrix Σ_f. (We could have extended the prediction step even further with an extra additive 'control-input' term representing the known impact of external driving variables $x(t)$ on the system state, which would still have left the KF analytically solvable.)

$$
\begin{aligned}
&\underline{\text{KALMAN FILTERING}}\text{ (multivariate state):}\\
&\text{Kalman Gain: } \mathbf{K} = \Sigma_p \mathbf{H}^\top (\mathbf{H}\Sigma_p\mathbf{H}^\top + \Sigma_y)^{-1}.\\
&\text{Data assimilation: } p[z_a|\mu_p, \Sigma_p, y, \Sigma_y] = N[\mu_a, \Sigma_a],\\
&\qquad \text{where } \mu_a = \mu_p + \mathbf{K}(y - \mathbf{H}\mu_p) \text{ and}\\
&\qquad \Sigma_a = (\mathbf{I} - \mathbf{K}\mathbf{H})\Sigma_p.\\
&\text{Prediction: } p[z_p|\mu_a, \Sigma_a] = N[\mu_p, \Sigma_p],\\
&\qquad \text{where } \mu_p = \mathbf{F}\mu_a \text{ and } \Sigma_p = \mathbf{F}\Sigma_a\mathbf{F}^\top + \Sigma_f.
\end{aligned}
\tag{22.4}
$$

The other major change in Eq. (22.4) compared to (22.3) is the *observational operator matrix* $\mathbf{H}$, which allows the observations to be a function of the states z rather than be the states themselves.

In KF, the product of observational operator and state estimate $\mathbf{H}z$ provides an estimate of the observations we expect to see when the system state is z. This is a very convenient component of the KF-procedure, because we often do not observe a system's state directly but must rely on measurement equipment that observes some derived property. An example would be where z consists of two components that differ in their temperature, but where we can only measure the overall average temperature. In that case, our observational operator would be a 1×2 matrix $\mathbf{H} = (0.5, 0.5)$ so that $\mathbf{H}z$ is the mean of the two temperatures $z[1]$ and $z[2]$. Time series models with such a separation between the true unobservable system state and observable properties that depend on the state are called *Hidden Markov Models* (HMM). So the KF is a HMM, although some people prefer to reserve the term HMM for time series models in which the system state space is discrete rather than continuous.

Let us now use the example of sensor-averaging to see a richer bivariate-state KF in action. Apart from z now being bivariate (and y measuring the average of the two), we use the same settings as for the simple univariate KF above. Our code for this matrix-version of Kalman Filtering is the following:

```
zp       <- Sp <- za <- Sa <- vector("list",n)
zp[[1]] <- c(0,0) ; Sp[[1]] <- Sf <- diag(1,2)
F        <- diag(0.9,2)
H        <- t( c(0.5,0.5) )
for(i in 1:n) {
  Ki        <- Sp[[i]] %*% t(H) %*% solve( H %*% Sp[[i]] %*% t(H) + Vy[i] )
  za[[i]]   <- zp[[i]] + Ki %*% ( y[i] - H %*% zp[[i]] )
  Sa[[i]]   <- ( diag(2) - Ki %*% H ) %*% Sp[[i]] ; if(i==n) break
  zp[[i+1]] <- F %*% za[[i]]
  Sp[[i+1]] <- F %*% Sa[[i]] %*% t(F) + Sf }
```

Note that our code above does not specify the observational uncertainties as covariance matrices because each $y(t)$ is a scalar as in our earlier example, so there is just a scalar observational variance at each time t. Also note that the code uses the R-data structure *lists* to collect the evolution of state estimates and uncertainty covariance matrices, which you can see from the use of the double square bracket notation [[]]. R-lists make it easy to address specific results. For example, `za[[30]]` is the final state vector estimate (after the last data assimilation), and `Sa[[30]]` is its uncertainty covariance matrix.

The results of this bivariate KF are shown in Fig. 22.3. Note that the figure shows the evolution of the mean and variance for our estimates of the *average* system state ($z(t) = (z_1(t) + z_2(t))/2$), not its two individual components. The average state's variances are thus calculated as $V(t) = (V_1(t) + V_2(t) + 2\,\mathrm{Cov}_{12}(t))\,/\,4$.

When we compare the results of the bivariate KF in Fig. 22.3 with what we found before for the univariate KF Fig. 22.2, there are clear differences, which may be surprising. The state estimates and the variances are all lower in the bivariate case. But the same information from data was used in both cases, with the same observational uncertainties and the same parameter settings. We also used the same

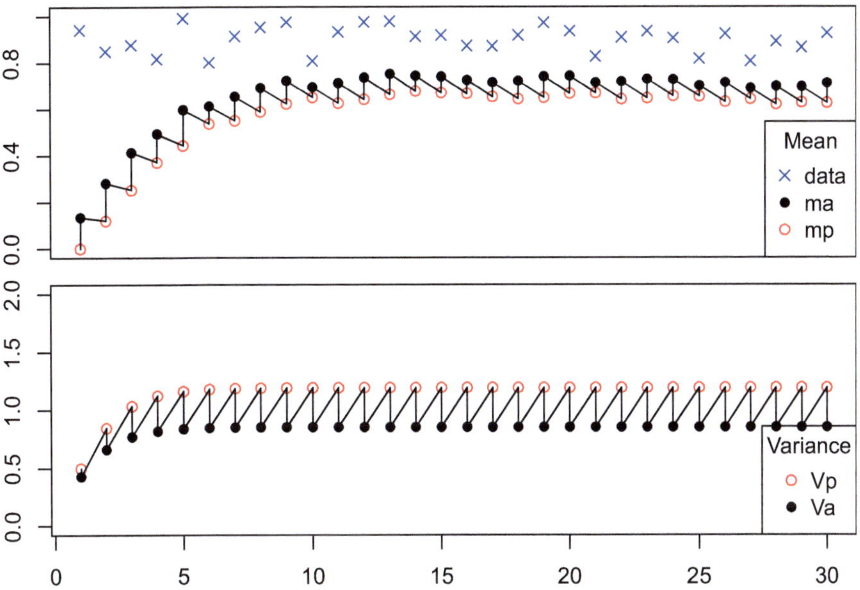

Fig. 22.3 Data assimilation using a bivariate KF. The top panel shows the observations and the alternating mean state estimates from model prediction (mp) and data assimilation (ma). The bottom panel shows the corresponding uncertainties. Observational uncertainty is not shown but has a constant variance equal to 3

model f that at each time step reduced the last assimilation estimate by 10% at a predictive uncertainty of σ_f^2 equal to 1 resp. Σ_f equal to the bivariate unit matrix. So why did the model f seem to have a greater pull, relative to the data, in the bivariate case? The answer is that we have introduced extra information by applying our model to the two halves of the state separately. If z_p has a variance of 1, then $z_p/2$ has a variance of only 1/4. So if our knowledge of the system is the same in both cases, then the values on the diagonal of Σ_f should be higher than σ_f^2. How much higher is a question that we leave as an exercise.

22.3 Time Series, KF and Complex Dynamic Models

We have seen that the simple linear time series models of this chapter can generate a large variety of system behaviour. And their calculation is efficient, even when we include sequential state calibration. These advantages have made the KF a widely used data assimilation procedure. The computationally most demanding part of KF is the matrix inversion in the calculation of the 'Kalman Gain' $\mathbf{K}$ (see Eq. (22.4)). The algorithm can be faster if we first calculate the *singular value decomposition*

(SVD) of $\mathbf{H}\Sigma_p\mathbf{H}^\top + \Sigma_y$ and then invert the SVD-components. We could code this in R as follows:

```
KalmanGain_SVD <- function(Sp,H,Vy) {
  Z      <- H %*% Sp %*% t(H) + Vy ; svd_Z <- svd(Z)
  inv_Z <- solve( t(svd_Z$v) ) %*% solve( diag(svd_Z$d) ) %*% solve( svd_Z$u )
  return( Sp %*% t(H) %*% inv_Z ) }
```

However, in the simple examples of this chapter, the SVD will not provide much computational advantage, nor will any other matrix decomposition.

The KF fits well in this book as its data assimilation step is Bayesian: the prior is the current predictive distribution for the state ($N[\mu_p, \Sigma_p]$), the likelihood is calculated from the difference of actual observations y and predicted observations ($\mathbf{H}\mu_p$) and the calculation of the posterior ($z_a|y \sim N[\mu_a, \Sigma_a]$) implicitly follows Bayes' Theorem without any need for MCMC because of the conjugate Gaussian prior-likelihood pair (see Chap. 5 for explanation of conjugacy).

However, the use of time series and sequential state corrections does have severe limitations. The first is that at each time step, there is no improvement to the model f. In data assimilation, we apply a correction to our state estimate in the direction of the observations, but we do not take away the need for such corrections, which may be that our model is poorly parameterised. Bayesian calibration of model *parameters* is lacking. We do show elsewhere in this book that the KF could be used for Bayesian calibration of parameters (of linear models), but it would not be the obvious method for that (Chap. 14).

Further, in KF, the model structure is constrained to be a state-space model where f is a linear transformation of the state effected by the matrix $\mathbf{F}$. So how can we apply the KF when we already have a reliable but nonlinear predictive model? An obvious solution to this is to *linearise* the model around the last state estimate. This is done by replacing the nonlinear function f with the so-called Jacobian matrix $\mathbf{J}_f$ of all first order partial derivatives: $\mathbf{J}_f[i, j] = \partial f_i/\partial z_j$. You can then use $\mathbf{J}_f$ in the calculation of the predictive variance: $\Sigma_p = \mathbf{J}_f\Sigma_a\mathbf{J}_f^\top + \Sigma_f$. For the calculation of the predictive mean you can of course still use the nonlinear function f itself: $z_p = f(z_a)$ (instead of $z_p = \mathbf{F}z_a$). Likewise, the observational operator matrix $\mathbf{H}$ can be a linearisation of a nonlinear function. A KF based on such linearisation is called an *Extended Kalman Filter* (EKF). In practice, the EKF may perform adequately if the time step at which correction data arrive is small, so we do not need to predict far into the future. In other words, linearisation is often acceptable as a *local* technique.

But what to do if our model f is so complex that we find it difficult to linearise, or when the time steps are too large? Our model may for example be a complex process-based model that predicts highly nonlinear system behaviour. We could then still apply the KF by creating an ensemble of multiple predictions from f (e.g. by sampling from its parameter distribution) and using the mean and variance of that set of predictions in the KF. That method is called *Ensemble Kalman Filtering* (EnKF) which is an example of the wider class of *particle filtering* methods. Of course, neither EKF nor EnKF addresses our concern that we want to improve our model structure and parameter values. Such data assimilation methods allow real-time correction for model discrepancy (Chap. 12) and poor parameterisation without ever learning to make better predictions in the first place.

Exercises

1. **Smoothness of a function represented by a GP**. What would you need to change in the second code-chunk to produce more smooth curves?
2. **Observational operator in KF.** How would you specify the observational operator **H** for a KF where the system state is trivariate with only the second (middle) variate being observable such that the measurement signal is 80% of the variate's true value?
3. **Bivariate vs. univariate KF.** What very small change do you need to make in the R-code for bivariate KF (which led to Fig. 22.3) such that the results become identical to those of the univariate KF in Fig. 22.2)?

Chapter 23
Spatial Modelling and Scaling Error

23.1 Spatial Models

Much of spatial modelling is a generalisation of the one-dimensional time series methods of the preceding chapter to two or more dimensions. Whereas time series methods discretise time, spatial models usually discretise space in the form of a raster of adjoining cells or a lattice of connected nodes. And Gaussian probability distributions feature equally prominently in both fields. An important class is that of *Gaussian Markov Random Fields* (GMRFs) which generalise AR(1) models to multiple dimensions, and which includes common spatial model types such as *Conditional Autoregressive* (CAR) and *Simultaneous Autoregressive* (SAR) models, which are characterised by different covariance functions. In AR(1) models, the system state $z(t)$ is determined by the value of the state one time step earlier. In GMRF, which are defined on a lattice, the state $z(s)$ is determined by the values in the *neighbourhood* of s, denoted as N_s. The neighbourhood consists of the nodes that have a direct link (or 'edge') to s. So if we want to predict the value of $z(s)$ at some point in our space of interest, and the values of all $s' \in N_s$ are known, then nodes further away can provide no additional information. We say that every $z(s)$ is *conditionally independent* from nodes outside its neighbourhood N_s. But note that if we were to write down the covariance matrix Σ for all points in our lattice, we would see a non-zero covariance for every pair of points irrespective of whether they are in each other's neighbourhood or not. We already saw that in the AR(1) models of the preceding chapter where every $z(t)$ was calculated from its immediate predecessor $z(t-1)$, but where the covariance function of the equivalent GP extended to all times. However, the *precision matrix* $\Omega = \Sigma^{-1}$ of our GMRF, i.e. the inverse of the covariance matrix, only has non-zero values on its diagonal and for those $\Omega[i, j]$ where s_i and s_j are in each other's neighbourhood. So the nodes and edges of the lattice structure tell us immediately where the non-zero values in Ω are. We already made use of this useful property of the precision matrix when we discussed graphical modelling (Chap. 16).

M. van Oijen, *Bayesian Compendium*,
https://doi.org/10.1007/978-3-031-66085-6_23

In the preceding chapter, we saw an example of an AR(1) time series model and an equivalent Gaussian Process (GP) model. The AR(1) was shown to be no more than a GP discretised on a sequence of times. Likewise, a GMRF is a GP discretised on a lattice. But GP models are in fact often employed in spatial modelling without discretisation, with *geostatistics* as the main example. Let's investigate that further.

23.2 Geostatistics Using a GP

The standard problem in geostatistical modelling is that of spatial prediction. We want to predict the values of $z(s)$ in some region, but we will receive measurement values only for some locations in that area. In that situation the following statistical assumptions are generally made. We assume that locations which are near to each other will more likely have similar values of z than points further away. Part of that similarity can be captured by a deterministic function $m(s)$ of the spatial coordinates (and possibly of covariates $x(s)$) across the whole area. But some of the similarity from proximity will be due to mechanisms in nature that we have little knowledge of and which cannot be represented by simple deterministic functions. We model that unexplained spatial process by means of a covariance function (or 'kernel function') $C(s, s')$. Finally, we allow for the possibility that there is some non-spatially correlated noise that makes $z(s)$ differ even from points $z(s + \delta)$ very nearby. That noise term is for historical reasons (associated with the mining industry) called a *nugget effect* and it is modelled as a zero-mean Gaussian. Putting everything together, this statistical modelling approach to geostatistics is a form of *Gaussian process regression* and can be written as (Diggle & Ribeiro, 2007):

GEOSTATISTICS USING A GAUSSIAN PROCESS:

$$z(s) = m(s) + T(s) + \epsilon(s), \text{ where } T(s) \sim GP[0, C(s, s')] \text{ and } \epsilon(s) \sim N[0, \sigma_\epsilon^2]. \tag{23.1}$$

This is a spatial linear Bayesian hierarchical model. It is of course almost the same statistical model as the one we studied in the chapter on model emulation (15). In that chapter our goal was to explore the output from a model f across its parameter space, whereas here we explore the state variable z across its physical space. The only difference is that we include the nugget-term $\epsilon(s)$ here. (But note that Andrianakis & Challenor, 2012 showed that even for model emulation the use of a GP with a nugget can make the required matrix inversions more robust.)

Let's develop the geostatistics approach using a concrete example for a rectangular region whose coordinates are between 0 and 4 in the east-west direction and between 0 and 3 north-south. We use a negative-exponential covariance function (as we did in our emulation examples) and assign values to the hyperparameters. At first we leave out the nugget, but we shall include it later.

(Note that we solved a simpler version of this same spatial problem in Chap. 16, Fig. 16.8. There we had no true mean function, so no regression parameters needed to be estimated.)

```
x1 <- c(0,4,4)     ; x2  <- c(3,0,3) ; y <- c(0.7,3.1,2.2) ; ny <- length(y)
Vy <- 1.47         ; phi <- 1.9115
x <- cbind(x1,x2) ; dx <- as.matrix( dist(x) ) ; Sy <- exp( -dx/phi ) * Vy
```

Because our model has no nugget yet, we can apply the same parameter estimation function GP.est that we defined in the chapter on GP and emulation (Chap. 15). When supplying the three data points defined in the preceding code-chunk, this gives us the following parameter estimates:

Geostatistical parameter estimation using a GP:

$$\mu_{\beta|y} = \begin{bmatrix} 1.2859 \\ 0.4238 \\ -0.2328 \end{bmatrix} ; \quad \Sigma_{\beta|y} = \begin{bmatrix} 2.885 & -0.469 & -0.591 \\ -0.469 & 0.14 & 0.065 \\ -0.591 & 0.065 & 0.224 \end{bmatrix}, \tag{23.2}$$

where the parameter vector is (intercept, slope west-east, slope north-south). Now let's see the predictive distribution for the point $s_0 = (0, 0)$. Deriving predictive distributions from a GP is also a problem we solved before in Chap. 15, and we use the R-function for prediction GP.pred that we developed there.

Geostatistical prediction for $s = (0, 0)$ using a GP:

$$p[z(s_0)|y] = N[\, 1.3181,\ 3.7413\,]. \tag{23.3}$$

23.3 Geostatistics Using geoR

Although in this book we prefer to use basic equations and basic R-code for our examples, to avoid the mysteries of black-box modelling, it is of course occasionally convenient to use R-packages, and geoR is a very convenient package for Bayesian geostatistics. We have used it before in Chap. 15, where it was shown to produce the same parameter estimates and predictions as basic code. Here we show the code to set up our above model in geoR, and then use the package to produce maps of estimates and their uncertainties for the whole region.

```
xy       <- as.geodata( cbind( x, y ) )
xpred1   <- x0[1] ; xpred2 <- x0[2] ; xpred.1 <- expand.grid( xpred1, xpred2 )
model.1 <- model.control( cov.m="exponential",
                          trend.d=~coords, trend.l=~coords )
prior.1 <- prior.control( beta.prior="normal", beta=mb, beta.var.std=Sb/Vy,
                          sigmasq.prior="fixed", sigmasq=Vy,
                          phi.prior="fixed", phi=phi,
                          tausq.rel.prior="fixed",tausq.rel=0 )
out.1    <- output.control(messages=FALSE)
geoR.1   <- krige.bayes( xy, loc=xpred.1, mod=model.1, pr=prior.1, out=out.1 )
```

Running this code and extracting the parameter estimates and variances from the geoR.1 object produce the same results as we found in Eqs. (23.2) and (23.3), which you can verify by running the geoR-code yourself. So we are reassured that we can use the package to plot the geostatistical predictions and predictive uncertainties (standard deviations) for the whole region, including our prediction point (0,0) and the three training points. The results are shown in Fig. 23.1. The calibrated mean function $m(s)$ is shown on the left, showing contour lines mainly running from southwest to northeast. The right panel shows the predictive uncertainties (square roots of the variances). At the three measured points the uncertainty has become zero, while increasing with greater distance from those points.

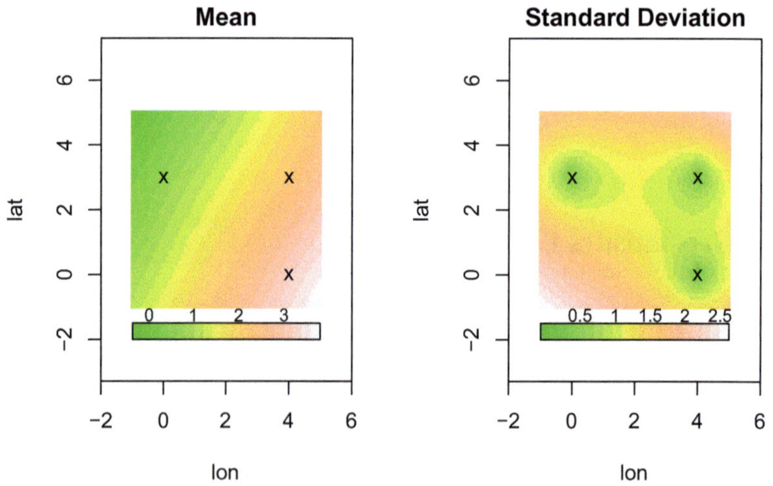

Fig. 23.1 Geostatistical prediction after calibration on three data points marked by crosses. GP without nugget. Left: predictive means; right: predictive standard deviations

23.4 Adding a Nugget

Adding a nugget of known variance is not difficult. In our basic equations, it amounts to adding a constant to all diagonal terms (= variances) of the spatial covariance matrix, without altering the off-diagonal covariances. And in geoR it is a matter of adding a non-zero relative nugget effect, denoted as tausq.rel in the package. We choose to add a nugget of 50% here. The only change required to the geoR-code to make this choice is in the definition of the prior distribution for parameters and hyperparameters:

```
prior.2 <- prior.control( beta.prior="normal", beta=mb, beta.var.std=Sb/Vy,
                          sigmasq.prior="fixed", sigmasq=Vy,
                          phi.prior="fixed", phi=phi,
                          tausq.rel.prior="fixed", tausq.rel=0.5 )
```

Running this leads to Fig. 23.2. We see that the introduction of the nugget has had little effect on the mean surface, but predictive uncertainties have increased. For example, the predictive standard deviation for the point (0, 0) increased from 1.934 to 2.242.

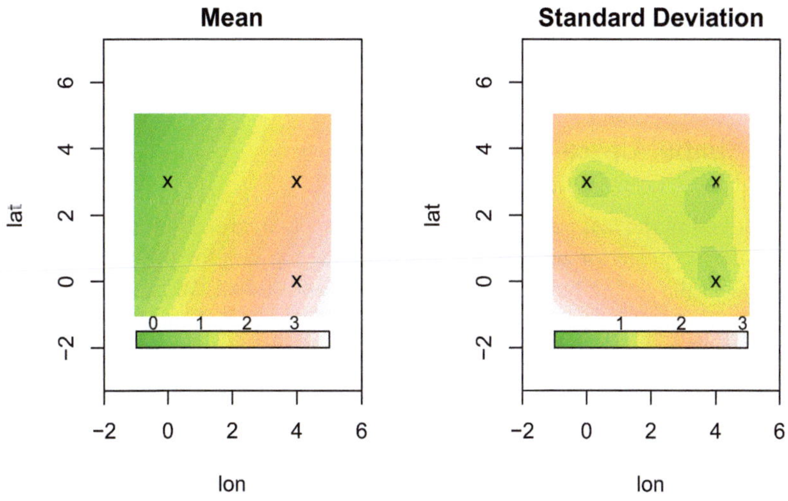

Fig. 23.2 Geostatistical prediction after calibration on three data points marked by crosses. GP with nugget. Left: predictive means; right: predictive standard deviations

23.5 Estimating All GP-hyperparameters

So far, we have only considered the case where the regression parameters β of the mean function of the GP are being estimated, with all other parameters considered to be known. Diggle et al. (2003) went a step further and showed an analytical solution for the case where β and a common scaling factor for Σ_β and Σ_y are estimated. Their analytical solution requires that the prior for the β is Gaussian, and for the variance-scaling factor is scaled inverse χ^2 (a form of inverse Gamma distribution). The marginal posterior distributions for β and for point prediction are then no longer Gaussian but more heavy-tailed t-distributions, reflecting the extra uncertainty that we have because of the unknown variances. Of course, a full Bayesian calibration would require all our uncertainties to be specified with prior distributions, but we quickly run out of analytical solutions if we try that. Diggle et al. therefore used MCMC to sample from the posterior distribution when also the correlation length ϕ was considered unknown, and that is how their package `geoR` operates when carrying out full Bayesian geostatistics. To speed up the computation, the package also requires that the domain of ϕ is discretised, so that the MCMC at least for that hyperparameter only needs to jump between a finite number of possible values. Note that such a full Bayesian spatial modelling approach requires an informative data set with observations at many more locations than the three we have in our simple example. For an example, see Levy et al. (2022), who refer to their spatial modelling as 'Bayesian kriging'.

23.6 Spatial Upscaling Error

As described in the last paragraph, the full Bayesian approach to geostatistics uses data to derive posterior probability distributions not just for the state variable at any point in the region but for all hyperparameters as well. We might be learning, for example, what the most probable value is for the spatial variance of some soil property $x(s)$. Spatial variability of $x(s)$ is of particular interest when x is an input to a model f, for example a process-based model. Let's make that more concrete with the following scenario that is very common in environmental modelling. We often want to use a model to make maps of $f(x)$ for some large region (country, continent, world), but have the problem that our model is slow to run. So we discretise space into many large grid cells, and for every grid cell we find the average value of x (which can be a scalar or vector), denoted as $E[x]$, and we calculate $f(E[x])$. Of course, we are then making an estimation error when f is nonlinear, because what we really should be calculating is $E[f(x)] = \int_{-\infty}^{+\infty} p[x]\, f(x)\, dx$, where $p[x]$ is the probability distribution that describes the variability of x in the cell. The erroneous assumption that $f(E[x]) = E[f(x)]$ is called the *fallacy of averages* and 'is perhaps the most widespread statistical error in biology' (Welsh et al., 1988). It is a scaling error: we are applying a model that is designed to have *point-support* (i.e. use input

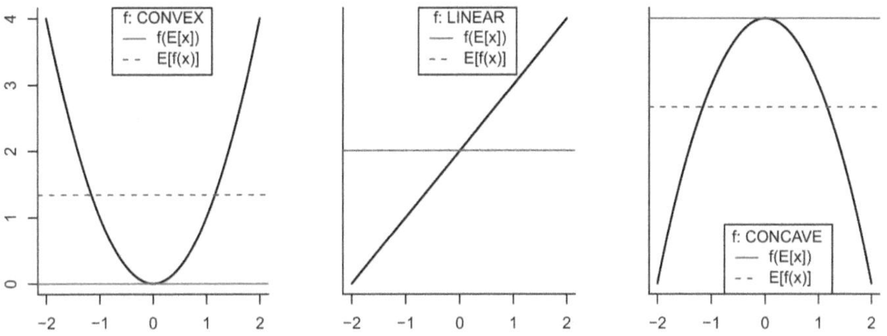

Fig. 23.3 Upscaling error for three functions f(x). The mean of f(x) in the interval [−2,2] is calculated incorrectly as f applied to the mean of x (solid horizontal red line) and correctly as the average of f(x) (dashed horizontal blue line). Upscaling error f(E[x]) − E[f(x)] is negative for convex functions, zero for linear functions, positive for concave functions

from a single point in space) to a large area. Let's denote the upscaling error as $\Delta = f(E[x]) - E[f(x)]$. Figure 23.3 shows for scalar x that Δ is negative for convex functions, zero for linear functions and positive for concave functions.

Fortunately, if we have a good estimate for the variance of x in the grid cell, then we can estimate the magnitude of the error from the Taylor expansion of f (Hines & Montgomery, 1990):

$$\begin{aligned} E[f(x)] &\approx f(E[x]) + \frac{1}{2}\mathrm{Var}[x] f^{(2)}(E[x]) \implies \\ \widehat{\Delta} &= -\frac{1}{2}\mathrm{Var}[x] f^{(2)}(E[x]), \end{aligned} \tag{23.4}$$

where $f^{(2)}(E[x])$ is the second derivative of f evaluated at the expectation value of x, and Var$[x]$ is the spatial variance of x. The formula tells us that (1) scaling error is expected to increase linearly with the variance of x, (2) its absolute value increases with the degree of nonlinearity of f (as measured by the magnitude of its second derivative), and (3) the sign of the scaling error (over- or underestimate) depends on the sign of the second derivative. That is quite a lot of information from such a simple equation!

Let's take the very simple example of a quadratic function to show how the formula works: $f(x) = x^2$ with x distributed uniformly in the unit interval [0,1]. So $E[x] = 0.5$, $\mathrm{Var}[x] = \frac{1}{12}$ and $f^{(2)}(E[x]) = 2$. Equation (23.4) then tells us to estimate the upscaling error as $\widehat{\Delta} = -\frac{1}{2}\frac{1}{12}2 = -\frac{1}{12}$. Now we need to verify that that is a good estimate. We find that $f(E[x]) = \frac{1}{2}^2 = \frac{1}{4}$ and $E[f(x)] = \int_0^1 x^2\,dx = \frac{1}{3}$, so $\Delta = \frac{1}{4} - \frac{1}{3} = -\frac{1}{12}$ which confirms that Eq. (23.4) did give us the correct error estimate for this particular function. Briefly returning to Fig. 23.3, which showed that the sign of the scaling error depends on the type of nonlinearity of the function, we see that Eq. (23.4) indeed estimates a negative Δ for functions with a positive second derivative, i.e. convex functions.

Equation (23.4) is formulated for scalar x but it generalises in a straightforward way to vectors (Levy et al., 2022; Van Oijen et al., 2017):

$$\widehat{\Delta} = -\frac{1}{2} tr(\Sigma_x \nabla_f^2), \tag{23.5}$$

where Σ_x is the covariance matrix of x, ∇_f^2 is the *Hessian matrix* of second-order partial derivatives of $f(x)$, and tr denotes the trace (sum of diagonal elements) of the matrix product $\Sigma_x \nabla_f^2$. This error estimation formula is suprisingly powerful. For several combinations of function shape and input probability distribution, the formula is exact, $\widehat{\Delta} = \Delta$, while in many others the formula provides a good approximation (Van Oijen et al., 2017). We tested the formula with three existing nonlinear models for emissions of greenhouse gases (methane, ammonia, nitrous oxide) across the UK and found UK-average upscaling errors for the three models to be -12%, -48% and -3%, in each case well estimated using the $\widehat{\Delta}$-formula. Of course, when the model f is not analytical, the Hessian matrix needs to be estimated numerically, and we provided R-code for the calculation of both the Hessian and $\widehat{\Delta}$ in the paper just mentioned.

Let's conclude. When we have used a fully Bayesian geostatistical approach to generate a spatial map of x, then we will also have acquired a posterior probability distribution for Var[x]. That uncertainty about spatial variance can be propagated forward using (23.4) to give us a probability distribution for the correction we need to make after we erroneously quantified the areal average of $f(x)$ as f applied to the areal average of x.

Exercise

1. **Scaling error.** Say an environmental variable x has a Gaussian distribution in a certain region: $x \sim N[\mu = 5, \sigma^2 = 10]$. We know that another variable z is a cubic function of x, so we can write $z = f(x) = x^3$. When we incorrectly estimate the regional mean value of z as $f(E[x]) = 5^3 = 125$, how big is the error that we make? Use Eq. (23.4).

Chapter 24
Spatio-Temporal Modelling and Adaptive Sampling

24.1 Spatio-Temporal Modelling

The previous chapter showed the similarity of models for time series and for spatial variation. So we should not expect that *spatio-temporal modelling* adds any completely new algorithms. Spatio-temporal modelling aims to estimate the changes over time of a spatially distributed dynamic system. We could actually use pure time series models $z(t)$ or spatial models $z(s)$ to achieve that, by defining z as location or time, respectively. So $z(t)$ could represent an individual's changing location over time t, and $z(s)$ could be the time at which a wildfire reaches location s. But the term 'spatio-temporal modelling' usually refers to models where the state variable z changes as a function of both spatial and temporal coordinates and is written as $z(s, t)$. Still, that does not require us to move to a new kind of model. We already saw that kernel-based approaches such as Gaussian Process modelling (GP) can be used for both spatial and temporal modelling, with the difference between the two mainly being one of dimensionality. A spatio-temporal system could certainly be modelled by means of a GP in three-dimensional space-time if we can define a sensible covariance function $C[(s, t), (s', t')]$. However, that is generally a difficult thing to do. It is very difficult to imagine what the proper covariance should be between our state variable z at location s and time t with the value of z at another place and time. Moroever, even if we could do so, this would lead to very large matrices that are difficult to invert and thus would cause computational problems for GP-estimation and -prediction. Therefore most spatio-temporal models employ *separable covariance functions* that can be decomposed as $C[(s, t), (s', t')] = C_s[s, s']C_t[t, t']$. The question of how similar we expect the system to be at nearby locations is thereby separated from the question of how we expect the system to evolve over time.

M. van Oijen, *Bayesian Compendium*,
https://doi.org/10.1007/978-3-031-66085-6_24

24.2 Adaptive Sampling

Separable spatio-temporal models can be very good tools to guide environmental monitoring. For example, if we want to keep track of the level of air pollution in a region, and only have a limited number of movable sensors available, then a spatio-temporal model can help us deploy the sensors at those locations which would most reduce our uncertainty about the regional average pollution level. Between measurements, we would be moving the sensors to new locations that minimise the model's predictive variance. In other words, the model would help us with *adaptive sampling* of the pollution levels across the region. We shall now look at a very simple example of such spatio-temporal modelling, where our region of interest is a 7×7 spatial grid, and where we follow the system over eight time steps. So both space and time are discretised. Our example is based on a very similar one provided by Wikle and Royle (1999), who in turn based their model on the separable spatio-temporal model of Huang and Cressie (1996) for melt-water prediction. The model uses an AR(1) model for time and the Kalman Filter (KF) equations for updating the spatial distribution, but we have already seen in the last two chapters that both are equivalent to GP-modelling. Initially, we set the temporal autoregression coefficient α to 0.9, and the spatial autoregression coefficient ρ to 0.75. Note that the spatial KF-calculation is for a 49-dimensional state variable z, because we have 49 different cells. We do not estimate any parameters or hyperparameters here, we just examine how sensor placement affects uncertainty about the state variable z.

We focus in this very simple example on changes in the *variance* of the state variable z, rather than on z itself. Given the simple model here (where all hyperparameters are assumed known), the variance is not informed by the data values y, only by the data locations and the variance for measurement uncertainty V_y. In fact, we arbitrarily choose the data values y to always be zero. What we want to minimise by sensible repositioning of a single roving sensor is the *Average Predictive Variance* (APV). So at each time step, our question is 'whereto should we now move our sensor such that the average variance over all 49 grid cells is minimal?'. We initialise our spatio-temporal model with R-code that you are familiar with from the preceding chapters:

```
coords   <- as.matrix( expand.grid(1:7,1:7) ) ; nz  <- dim(coords)[1]
tausq    <- 10 ; Vy <- diag( tausq, 1 ) ; ny <- 8 ; y <- 0
mb       <-  0 ; Va <- 80 ; rho <- 0.75
Rspatial <- rho^as.matrix( dist(coords) )
zp       <- vector("list",ny)   ; Sp       <- vector("list",ny)
za       <- vector("list",ny+1) ; Sa       <- vector("list",ny+1)
za[[1]]  <- rep( mb, nz )       ; Sa[[1]] <- Va * Rspatial
APV      <- rep(NA,ny+1)        ; APV[1]  <- mean( diag( Sa[[1]] ) )
alpha    <- 0.9                 ; M <- diag(alpha,nz)
Vf       <- 20                  ; Q <- Vf * Rspatial
iSensor  <- rep(NA,ny)
```

After this initialisation, our spatio-temporal algorithm for adaptive sampling will cycle through the following four steps:

1. FORECAST: using the model to propagate `za[[i]]`, `Sa[[i]]` to `zp[[i]]`, `Sp[[i]]`.
2. DESIGN: selecting the sensor placement that gives lowest APV among all 49 possible locations.
3. OBSERVATION: using the selected sensor location to measure y (this step can be skipped as long as we are only interested in the magnitude of predictive uncertainty).
4. ANALYSIS: using the sensor information (in particular its location) to update predictive variance.

We code this cycle in R as follows:

```
KalmanGain <- function(Sp,H,Vy){ Sp %*% t(H) %*% solve(H%*%Sp%*%t(H) + Vy) }
for (i in 1:ny) {
  zp[[i]] <- M %*% za[[i]] ; Sp[[i]] <- M %*% Sa[[i]] %*% t(M) + Q
  r_APV <- rep(NA,nz)
  for (r in 1:nz) {
    r_Hi      <- matrix(0,nrow=1,ncol=nz) ; r_Hi[r] <- 1
    r_Ki      <- KalmanGain( Sp[[i]], r_Hi, Vy )
    r_Sa      <- ( diag(nz) - r_Ki %*% r_Hi ) %*% Sp[[i]]
    r_APV[r] <- mean( diag(r_Sa) ) }
  iSensor[i] <- match( min(r_APV), r_APV )
  Hi         <- matrix(0,nrow=1,ncol=nz) ; Hi[iSensor[i]] <- 1
  Ki         <- KalmanGain( Sp[[i]], Hi, Vy )
  za[[i+1]]  <- zp[[i]] + Ki %*% ( y - Hi %*% zp[[i]] )
  Sa[[i+1]]  <- ( diag(nz) - Ki %*% Hi ) %*% Sp[[i]]
  APV[i+1]   <- mean( diag( Sa[[i+1]] ) ) }
```

The results of this exercise are shown in Fig. 24.1. We see that the algorithm initially places the sensor in the exact centre, from where it reduces uncertainty as measured by the predictive variance in all directions because of the autoregressive spatial covariance function. The uncertainty is of course lowest at the sensor position itself. Over the following steps the sensor keeps exploring all four quadrants. Because this spatio-temporal system has memory (quantified by the temporal autoregression coefficient α), uncertainty stays reduced at locations where the sensor was placed in the recent past. So how would the sensor be moving around if the spatial correlation (ρ) or memory strength (α) were changed? You can explore that yourself by running the code, but here are some results:

- If we set $\rho = 1$, knowledge of one cell gives perfect information about all others at the same time, so the sensor does not move at all, and uncertainties are very low.
- If we set $\rho = 0$, knowledge of one cell gives no information at all about other cells, so the algorithm starts in a corner and then always moves to the nearest not-yet-used location.
- If we set $\alpha < 0.3$, memory is limited, and the sensor stays at the central location.

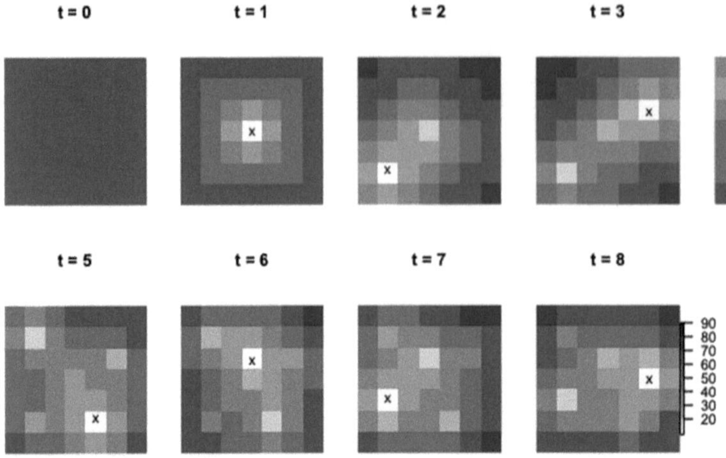

Fig. 24.1 Adaptive sampling by a single roving sensor in a region of 7×7 grid cells. Cell colours indicate predictive variance from low (white) to high (black). Current sensor position is marked by a cross

All this makes intuitive sense, which suggests that such an adaptive sampling algorithm can be useful for managing monitoring campaigns.

24.3 Extensions and Applications

Our adaptive sampling example can easily be extended to include parameter estimation, which would make the actual measurement values y informative. That can be done by extending the state vector with entries for the parameters. We already gave an example of using the KF for parameter estimation when we compared methods for Bayesian estimation of linear models (Chap. 14). We could even consider expanding the Bayesian learning during adaptive sampling to include learning which spatio-temporal *model* is to be preferred! To choose between different spatial covariance functions, for example, we could start with a Bayes Factor equal to one before the sampling has commenced. That approach would be comparable to adaptive sampling algorithms used in MCMC.

Adaptive sampling can be used for the development of emulators too. That implies optimising locations in parameter space (where to evaluate the model next) rather than real space (where to measure next). *Bayesian optimisation* of the hyperparameters of machine learning algorithms is an example of this. It is an iterative procedure whereby a likelihood-emulating GP is optimised using adaptive sampling in hyperparameter space (see Chap. 21).

In our example we assumed that our quantity of interest was the average predictive variance. In practical problems, with complicated logistics and costs of sensor movement, maintenance and repair, the quantity of interest may deserve to be chosen more carefully. We could use Bayesian decision theory (Chap. 19) to assess the trade-off between costs (time and money) and gains (reduction of uncertainty).

For several more recent examples of Bayesian GP-assisted sampling algorithms, see Xu et al. (2016).

Exercise

1. **Explore adaptive sampling.** Explore how varying the parameters ρ and α controls adaptive sampling, as suggested in this chapter.

Chapter 25
What Next?

We are nearing the end of this book, and it is time to take stock. How far have we come and what can we do next?

I hope that this book has made clear why Bayesian methods are increasingly being used by scientists. The approach offers a consistent way of quantifying and reducing uncertainties, rigorously based on probability theory. The concepts are simple, consisting of no more than defining and combining priors and likelihoods. It provides us with a straightforward procedure for using data and assessing the reliability of models and model predictions.

The Bayesian outlook also helps us to see the forest for the trees in computational statistics. There is an enormous amount of confusing terminology, and we can choose from a very large number of different statistical methods with impressive-sounding names. But under the hood the differences are often small. Many statistical models are based on the same underlying mathematics and statistical assumptions, as we have mentioned in several chapters. Machine learning is just nonlinear regression, geostatistics and emulation of PBMs are both applications of Gaussian processes, graphical modelling and hierarchical modelling are representations of joint probability distributions. Kalman Filtering, which is commonly presented as a form of recurrent state estimation, can also be used for estimating parameter values.

Much of the applied statistics literature is about novel ways to linearise, Gaussianise (the term 'normalise' is already taken) and discretise complexities away such that we can keep using standard techniques. It is all in the service of providing computationally efficient and sufficiently accurate estimates with well-quantified uncertainty. But even where methods are truly new, we can always try to understand them—and see how they are related—by asking the Bayesian questions: What is the statistical model? What is the stated or implied prior distribution? What is the likelihood function? How is Bayes' Theorem (explicitly or implicitly) used to find the posterior distribution? How is the posterior distribution used to assess the significance of the results?

M. van Oijen, *Bayesian Compendium*,
https://doi.org/10.1007/978-3-031-66085-6_25

I hope that seeing the family relationships between all kinds of different statistical methods has helped you get an overview of the field, and an understanding of the generality of Bayesian thinking. Jaynes used to refer to probability theory as "the logic of science", but it is of course everyone's logic. There is no difference in thinking between statisticians and scientists.

25.1 Some Crystal Ball Gazing

The following is my perspective on the future of Bayesian methodology, which, as the result of using a crystal ball, comes with considerable uncertainty. Some of the ideas were already mentioned by Van Oijen (2017).

Bayesian methods are computationally demanding and their recent uptake by many modellers has only been made possible by the advent of fast computational methods for sampling from distributions, particularly MCMC. This has stimulated interest in the application of Bayesian methods to complex, slow models and to complex uncertainty assessments where multiple models are considered or where parameter uncertainty is defined hierarchically. But in such cases, computational demand becomes a bottleneck again, and we expect that the search for new, faster MCMC-algorithms and reliable approximation methods will continue. This will include attempts to develop better (faster and more reliable) model emulators, and there we may well see an increased role of machine learning. A separate line of research will be efforts to design mechanistic rather than statistical emulators using ideas from the field of systems identification.

When we develop faster methods for process-based modelling (PBM), it will also become possible to embed them in Bayesian hierarchical models (BHM) (Fer et al., 2021). That has so far not been possible because of the computational demand. Bayesian calibration of a parameter-rich and slow process-based model is already hard, and allowing its parameters to vary in space or within a population exacerbates the problem. However, for scientists, a generally applicable approach to 'BHM-PBM' would constitute the holy grail as it would combine mechanistic understanding of the system (in the PBM) with knowledge of variability (in the BHM).

Another important issue that will remain is how to include the discrepancy between model behaviour and real system dynamics in the Bayesian analysis. We expect intensified methodological research on how to represent discrepancy and prior uncertainty about it. Ensemble modelling, which addresses model structural uncertainty too, is also likely to remain of key interest.

Besides *quantifying* model discrepancy, Bayesian methods will also be increasingly used to *reduce* discrepancy, i.e. improve model structure. Bayesian calibration, by reducing uncertainty about model parameter values, focuses our attention on the remaining uncertainty about model structural error. Machine learning methods help pinpoint such errors when they provide better predictions for system behaviour than current PBMs do, which may be the conclusion from Bayesian model comparison.

Whenever a machine learning model is superior to a PBM, our understanding and model implementation of causal mechanisms in the latter is shown to be inadequate.

Uncertainties that have been quantified and reduced using Bayesian methods must be communicated to end-users of the modelling studies, but there is no standard approach for reporting uncertainties. However, research on uncertainty communication is continuing and graphical modelling may play an increasingly prominent role. Graphical models have, for example, been used to convey uncertainty about the impact of management decisions on the different services that a single ecosystem can provide (e.g. Smith et al., 2012).

These issues may acquire even more significance in the future with the expected increase in frequency and intensity of extreme climatic events, and their concomitant risks (Reyer, 2015). Risk analysis is a form of uncertainty analysis and will therefore benefit from Bayesian methods to provide reliable estimates of risk, vulnerability and hazard probability. But risk analysis as practiced is often informal; it needs to be formulated more rigorously and embedded within Bayesian decision theory.

Finally, and in a very general sense, I believe that there will be a convergence or linking of methodology that we use for explanation (such as process-based models), prediction (such as machine learning) and uncertainty quantification (Bayes) (Bradley et al., 2022). The convergence will be stimulated by common goals, such as identifying causal pathways from an ever-increasing stream of data.

25.2 Further Reading

I hope that this book has stimulated your interest in Bayesian methods, and in probabilistic thinking more generally. If you want to read more, there are of course thousands of Bayesian research papers to study. There are also many good books, and I can only mention a few of them. The following four books have been written by and for scientists: *Probability theory: the logic of science* by Jaynes (2003), *Bayesian Logical Data Analysis for the Physical Sciences* by Gregory (2005), *Data Analysis: A Bayesian Tutorial* by Sivia and Skilling (2006) and *Practical Bayesian Inference: A Primer for Physical Scientists* by Bailer-Jones (2017). The book by Jaynes is unique in its clarity and depth. It does not discuss many statistical models, nor any computational algorithms—it is a book about how to think. It demonstrates how Bayesian thinking resolves many paradoxes of the classical statistical literature. The books by Gregory, Sivia and Bailer-Jones do discuss some concrete ways of analysing data. They focus on topics that are covered in the first half of this book, with Gregory using examples in the computer-algebra system Mathematica, Bailer-Jones using R, and Sivia staying with (very readable) mathematical exposition. A stand-out introductory text is *Statistical Rethinking* by McElreath (2016). He uses code-examples in R and Stan but his English-language explanations in particular form an impressively clear and jargon-free introduction to Bayes. This does come at the cost of requiring many pages of text to convey the ideas. *Bayesian Data Analysis* by Gelman et al. (2013) is the bible of Bayesian statisticians. It is not aimed at scientists, but it is where you may want to look up how different statistical models work, their underlying assumptions as well as their strengths and weaknesses.

25.3 Closing Words

The main message of this book is that Bayesian methods give you complete freedom in data analysis and modelling. You can choose any probability distribution that adequately represents your state of knowledge, not restricting yourself to Gaussians for mathematical convenience. Bayes' Theorem always applies. And the advent of sampling-based Bayesian calibration means that you can choose to work with realistic nonlinear models, such as PBMs. You need not rely on the methods that statisticians had to invent, and the assumptions that they had to make, when computing power was limited and models needed to be solved by hand. I hope that you will go out and enjoy the freedom that Bayesian methods coupled with computing power give the modern scientist!

Appendix A
Notation and Abbreviations

Notation

The two main conventions that we follow are:

1. Square brackets are used with probabilities, probability distributions, expectations and likelihoods: $p[\theta|y]$, $N[\mu, \sigma^2]$, $E[x]$, $L[\theta]$ etc.
2. Parentheses are used for functions, such as $f(x, \theta)$.

   The most commonly used symbols are:

	Mathematics	R-code
Likelihood	$L[\theta]$, $p[y\|\theta]$	`L`
Variance (scalar); Covariance matrix	Var[ ] or σ^2; Σ	`V, S`
Precision matrix	W	`W`
Covariates, predictor variables	x	`x`
Design matrix	**X**	`X`
Observations	y	`y`
True value or model output	z	`z`
Parameter (scalar or vector)	θ, β	`b`
Maximum likelihood estimate	θ^{MLE}	-
Prior mean and covariance matrix	μ_β, Σ_β	`mb, Sb`
Posterior mean and covariance matrix	$\mu_{\beta\|y}$, $\Sigma_{\beta\|y}$	`mb_y, Sb_y`
Covariance matrix for measurement uncertainty	Σ_y	`Sy`
Correlation length	ϕ	`phi`

M. van Oijen, *Bayesian Compendium*,
https://doi.org/10.1007/978-3-031-66085-6

Abbreviations

AR	Autoregressive time series model
A-R	Accept-Reject (or Rejection Sampling)
BC	Bayesian Calibration
BDT	Bayesian Decision Theory
BHM	Bayesian Hierarchical Model
BMA	Bayesian Model Averaging
BMC	Bayesian Model Comparison
DAG	Directed Acyclic Graph
EXPOL5, EXPOL6	5- and 6-parameter version of the expolinear model
GAM	Generalised Additive Model
GAMM	Generalised Additive Mixed Model
GBN	Gaussian Bayesian Network (or Gaussian Belief Network)
geoR	R-package for Bayesian geostatistics
GLM	Generalised Linear Model
GLMM	Generalised Linear Mixed Model
GLS	Generalised Least Squares
GM	Graphical Model
GP	Gaussian Process
JAGS	Just Another Gibbs Sampler (software for MCMC)
KF	Kalman Filter
LMM	Linear Mixed Model
MCMC	Markov Chain Monte Carlo
PBM	Process-Based Model
PRA	Probabilistic Risk Analysis

Appendix B
Mathematics for Modellers

This book is not very mathematical, but here are some elementary pointers.

How to Read an Equation

There are three ways to check equations:

1. The scientist's way: check dimensions,
2. The mathematician's way: replace variables with 0, 1 or ∞ (infinity),
3. Everyone's way: make a plot.

Let's take the example of the following equation:

$$p[\theta] = \frac{1}{\sigma\sqrt{2\pi}} \exp\left(-\frac{(\theta-\mu)^2}{2\sigma^2}\right).$$

It defines the function $p[\theta]$. Now we apply our three checks:

1. θ and μ must be in the same units because they are subtracted from each other. And σ must be in the same units too, because exponentiated quantities (everything between the brackets of the exp-function) must be dimensionless. That means that $p[\theta]$ has the inverse units of θ itself. That is as far as the dimensional analysis can take you for this equation: your physical knowledge of θ should now take over. (If you happen to know that the equation defines a probability density for θ, then you will not be surprised about $p[\theta]$ having the inverse units of θ.)
2. After various replacements with 0, 1 or ∞, you will find that:
   - the function goes to zero when θ or μ goes to plus or minus infinity,

M. van Oijen, *Bayesian Compendium*,
https://doi.org/10.1007/978-3-031-66085-6

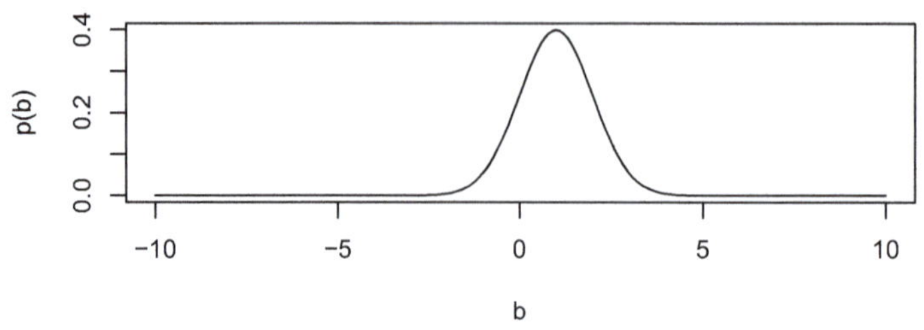

Fig. B.1 A function.

- if θ and μ are both replaced by the same constant, then the function reaches its maximum value,
- σ cannot be zero because that would give a division by zero.
- a very large value for σ would make $p[\theta]$ close to zero everywhere.

3. Plotting gives an even better idea of the shape of the function. Let's replace μ and σ by 1, and plot the function. Here is R-code and the result is in Fig. B.1:

```
p <- function(b,m=1,s=1) {exp( -0.5 * ((b-m)/s)^2 ) / (s*sqrt(2*pi))}
b <- seq(-10,10,by=0.1) ; plot( b, p(b), type="l" )
```

The figure confirms what we already had found out about the behaviour of $p[\theta]$, and the R-function `p(b, m, s)` that we wrote allows us to explore the function for different values of μ and σ.

Dimension Checking of Linear Algebra

There is some linear algebra in this book because that makes it easy to work with multivariate quantities and apply linear transformations to them. Take the following equation from Chap. 14:

$$\mu_{\beta|y} = \mu_\beta + \mathbf{K}(y - \mathbf{X}\mu_\beta).$$

Here $\mu_{\beta|y}$, μ_β and y are vectors and $\mathbf{K}$ and $\mathbf{X}$ are matrices. The equation defines how each element in μ_β is transformed into the corresponding element of $\mu_{\beta|y}$, so we are effectively writing multiple equations in one go.

Linear algebra comes with its own *mathematical* dimension check. Let's say that the dimensions of $\mathbf{K}$ are 2×3 (number of rows times number of columns: the rows are always given first). Knowing the dimensions of $\mathbf{K}$ means that the dimensions of all other quantities in the equation are determined too! This is because of how multiplication in linear algebra works. Say we have a vector or matrix A with dimensions ($r_A \times c_A$), and a vector or matrix B that is ($r_B \times c_B$). Now, we can only multiply them as $C = AB$ if $c_A = r_B$ and the dimensions of C will then be ($r_A \times c_B$). If we now go back to our equation where **K** is (2×3), we find that the μ-vectors must be (2×1), the y-vector (3×1) and the matrix **X** (3×2), as you can verify for yourself.

Appendix C
Probability Theory for Modellers

Notation

- $p[x]$ = Probability of x. We use this notation not only for discrete distributions but also for continuous ones where we should strictly speak of the 'probability density' of x.
- $p[x|A]$ = Conditional probability of x given that A is true.
- $\neg x$ = Negation of x. For example, $p[\neg x]$ is the probability of x not being true.

Probability Distributions

A probability distribution is a non-negative function that sums to 1 (if it has a discrete domain) or integrates to 1 (if its domain is continuous). The domain can be univariate or multivariate. You can define your own probability distribution if you want—no need to rely on distributions that already have a name.

Product Rule of Probability

$$p[A \cap B] = p[A]\, p[B|A] = p[B]\, p[A|B], \tag{C.1}$$

where $p[A \cap B]$ stands for the probability that both A and B are true.

M. van Oijen, *Bayesian Compendium*,
https://doi.org/10.1007/978-3-031-66085-6

Law of Total Probability

If $p[\theta]$ obeys a discrete distribution, then

$$p[y] = p[\theta]\, p[y|\theta] + p[\neg\theta]\, p[y|\neg\theta] = \sum_i p[\theta_i]\, p[y|\theta_i]. \tag{C.2}$$

If $p[\theta]$ obeys a continuous distribution, then

$$p[y] = \int p[\theta]\, p[y|\theta]\, d\theta. \tag{C.3}$$

Bayes' Theorem

$$p[\theta|y] = \frac{p[\theta]\, p[y|\theta]}{p[y]} \tag{C.4}$$

Sequential Bayesian Updating

Bayes' Theorem can be used to calculate how a dataset changes our prior probability distribution for parameter(s) $p[\theta]$. Say that we receive a dataset y_1 and have calculated the posterior distribution $p[\theta|y_1]$. Now assume that later we receive another, independent dataset y_2, and want to calculate a posterior distribution that incorporates the information from both datasets, so $p[\theta|y_2, y_1]$. We can obviously do that by starting again at the beginning, with $p[\theta]$, and feeding in both datasets at the same time into Bayes' Theorem, rather than just y_1. But it may be more efficient if we take the posterior distribution from using the first dataset, i.e. $p[\theta|y_1]$, and consider that to be the prior distribution for a second application of Bayes' Theorem. This is a valid approach as we can show:

$$\begin{aligned} p[\theta|y_2, y_1] &= \frac{p[\theta|y_1]\, p[y_2|\theta, y_1]}{p[y_2|y_1]} \\ &\propto p[\theta|y_1]\, p[y_2|\theta] \\ &\propto p[\theta]\, p[y_1|\theta]\, p[y_2|\theta]. \end{aligned} \tag{C.5}$$

The simplification in the second line is possible because y_1 and y_2 are supposed to be independent. In the last line, we expanded using Bayes' Theorem with the first dataset. Note that the likelihood calculation in the last line consists simply of multiplying the likelihood functions for the two datasets. That shows that the two-step procedure is valid and gives the same result as the one step procedure.

Gaussian Probability Distributions

If a scalar parameter θ has a Gaussian (or 'normal') distribution with mean μ and variance σ^2, we can write that succinctly as $\theta \sim N[\mu, \sigma^2]$. We can also give the formula of the probability distribution for θ:

$$p[\theta] = \frac{1}{\sigma\sqrt{2\pi}} \exp\left(-\frac{(\theta-\mu)^2}{2\sigma^2}\right). \tag{C.6}$$

If a parameter vector θ (dimension $n \times 1$) has a multivariate Gaussian distribution with mean vector μ and covariance matrix Σ, we can write that succinctly as $\theta \sim N[\mu, \Sigma]$. The formula for the multivariate Gaussian is easiest written with matrix notation. But we first rewrite the formula for the univariate Gaussian in such a way that the similarity with the multivariate Gaussian will be most apparent when we write them underneath each other:

- Univariate Gaussian distribution:

$$p[\theta] = (2\pi\sigma^2)^{-1/2} \exp\left(-\frac{1}{2}(\theta-\mu)(\sigma^2)^{-1}(\theta-\mu)\right). \tag{C.6'}$$

- Multivariate Gaussian distribution:

$$p[\theta] = |2\pi\Sigma|^{-1/2} \exp\left(-\frac{1}{2}(\theta-\mu)^\top\Sigma^{-1}(\theta-\mu)\right). \tag{C.7}$$

A nice property of the multivariate Gaussian is that conditional probabilities, where some of the variates are exactly known, can easily be calculated. In geostatistics, for example, we often set up a multivariate Gaussian $p[z] = N[\mu, \Sigma]$ for a vector z of dimension $n \times 1$, where one variate z_0 is unknown and the remaining $n-1$ are known. A probabilistic prediction for the single unknown value is then written as $p[z_0|z_1]$, where z_1 is the known sub-vector of dimension $(n-1 \times 1)$. Before writing out the general formula for a conditional multivariate Gaussian, we first define five symbols.

- $z = \begin{pmatrix} z_0 \\ z_1 \end{pmatrix}$, $\mu = \begin{pmatrix} \mu_0 \\ \mu_1 \end{pmatrix}$, $\Sigma = \begin{pmatrix} \Sigma_{00} & \Sigma_{01} \\ \Sigma_{10} & \Sigma_{11} \end{pmatrix}$,
- $m_0 = \mu_0 + \Sigma_{01}\Sigma_{11}^{-1}(z_1 - \mu_1)$,
- $V_0 = \Sigma_{00} - \Sigma_{01}\Sigma_{11}^{-1}\Sigma_{10}$,

where $\{z_0, \mu_0, \Sigma_{00}, V_0\}$ are scalars with dimension (1×1), $\{z_1, \mu_1, \Sigma_{10}, \Sigma_{01}^\top\}$ are $(n-1 \times 1)$ vectors, and Σ_{11} is an $(n-1 \times n-1)$ matrix. With these definitions, the conditional probability for the multivariate Gaussian is:

$$p[z_0|z_1] = |2\pi V_0|^{-1/2} \exp\left(-\frac{1}{2}(z_0 - m_0)^\top V_0^{-1}(z_0 - m_0)\right). \tag{C.8}$$

In fact, the formula is still correct if z_0 is not a scalar but multivariate, but the partitioning into one unknown and $n-1$ unknowns occurs more often.

Appendix D
R

R is a freely available computer language for statistical and scientific computing that keeps developing. There is a large and active user community that produces new functionality in the form of *packages* that can be downloaded and installed when needed. Here we only introduce those parts of R that are used in this book. For scientists who desire a comprehensive introduction to R, a great resource is the online book *R for Data Science* by Hadley Wickham and Garrett Grolemund: https://r4ds.had.co.nz. *RStudio* (https://rstudio.com) is an integrated development environment that makes working with R a joy, and it facilitates literate programming by means of Yihui Xie's *R Markdown* and *bookdown* (https://bookdown.org).

Basic R Commands

Variables can be scalars, vectors, matrices:

```
x    <- 2
y    <- c(1,2)
yrow <- matrix( c(1,2), nrow=1) ; ycol <- matrix( c(1,2), ncol=1)
Z    <- matrix( 0:3, nrow=2)
```

Note the three different vector definitions. `y` is neither a row- nor a column-vector. R will try to interpret its intended meaning from how it is used in the code. `yrow` is a row-vector and `ycol` is a column-vector. We can check the dimensions of variables using the `length`() and `dim`() functions. `length`() is defined for all variables, `dim`() only for those where the numbers of rows and columns are known.

M. van Oijen, *Bayesian Compendium*,
https://doi.org/10.1007/978-3-031-66085-6

We can use the `print` command to show the values of variables on our screen:

```
  print( Z ) ; print( length(Z) ) ; print( dim(Z) )
>      [,1] [,2]
> [1,]    0    2
> [2,]    1    3
> [1] 4
> [1] 2 2
```

Ordinary multiplication uses the '$*$' operator, matrix multiplication uses '$\%*\%$'.

```
xZ <- x * Z
Zy <- Z %*% y ; yZ <- y %*% Z ; Zyc <- Z %*% ycol ; yrZ <- yrow %*% Z
```

Diagonal matrices can be specified with the `diag` function:

```
I5a <- diag( 5 ) ; I5b <- diag( rep(1,5) ) ; X <- diag( 1:3 )
```

The inverse and transpose of matrices are calculated with the `solve` and `t` functions:

```
invZ <- solve(Z) ; trZ <- t(Z)
```

R has a *list* data structure, which can combine variables of any type. Elements from the list can be addressed by list-name + $ + element-name, or by using square brackets:

```
  mylist <- list( distr="Bivariate Gaussian", m=c(0,0), S=diag(2) )
  print( mylist$distr ) ; print( mylist[ 2:3 ] )
> [1] "Bivariate Gaussian"
> $m
> [1] 0 0
>
> $S
>      [,1] [,2]
> [1,]    1    0
> [2,]    0    1
```

You can define your own functions, say a quadratic:

```
myf <- function(x){ x^2 }
```

Basic plotting is done with the `curve` and `plot` commands (the latter allows more detailed formatting):

```
curve( myf )
plot( 1:10, myf(1:10), main="My function", xlab="x", ylab="y" )
```

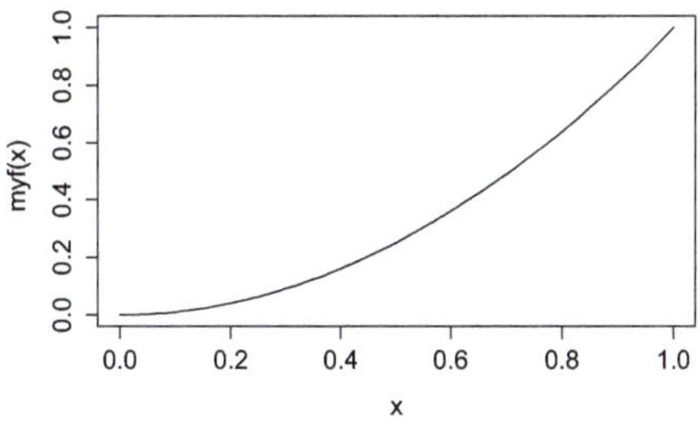

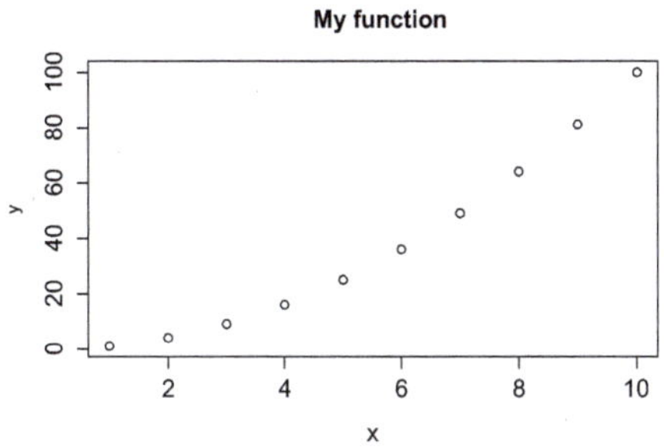

R has many features for working with probability distributions. Random numbers are generated with **runif**, **rnorm**, etc. Probability densities are derived with functions **dunif**, **dnorm**, etc.

```
x  <- rnorm( 1e4, mean=1, sd=0.3 )
dx <- dnorm( x  , mean=1, sd=0.3 )
```

For continuous probability distributions, use **hist** to plot a sample histogram, or **curve** to plot the density function:

```
hist ( x )
curve( dnorm,-2,2 )
curve( dbeta(x,4,2) )
```

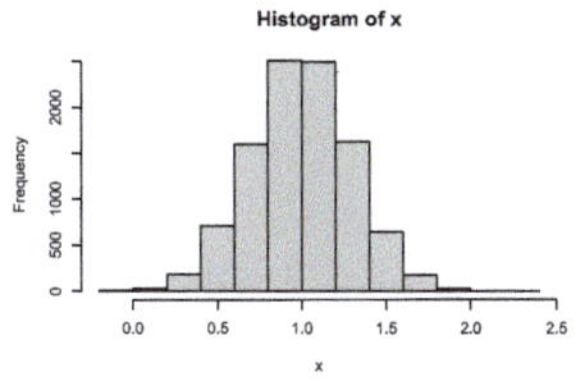

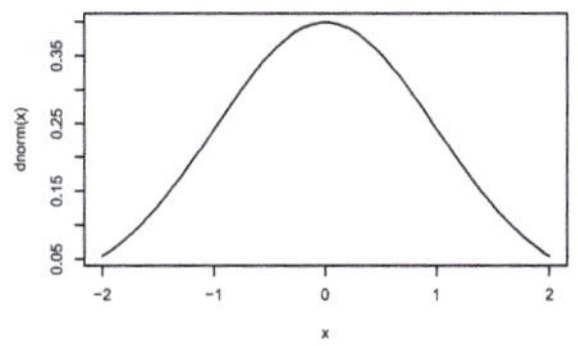

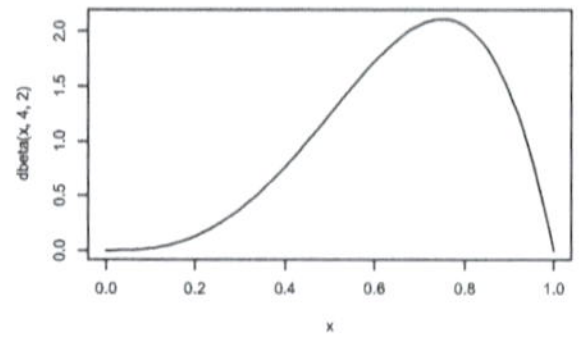

In the case of discrete probability distributions, use `barplot`:

```
barplot( dbinom( 0: 1, size= 1, p=0.7 ), names.arg=0: 1, main="Br[x]" )
barplot( dbinom( 0:10, size=10, p=0.7 ), names.arg=0:10, main="Bi[x]" )
barplot( dpois ( 0:10, lambda=1 )       , names.arg=0:10, main="Pn[x]" )
```

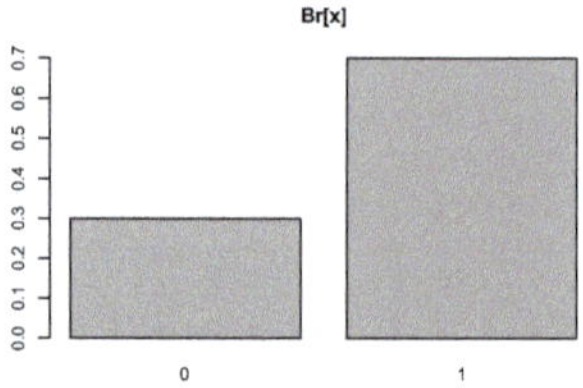

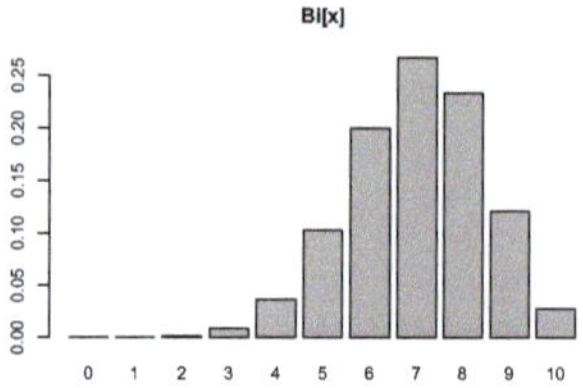

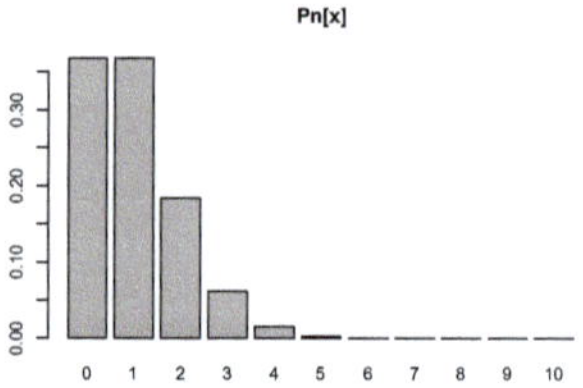

R-packages

R-packages expand the functionality of R. You can find most of them on https://cran.r-project.org. R-packages need only be installed once on your computer, using the `install.packages()` command. Thereafter you can load them whenever you need them with the `library()` command. Take the example of the `mvtnorm` package which contains useful functions for multivariate Gaussian distributions:

```
sample <- rmvnorm( 100, sigma=diag(1,2) ) # Gives error message
install.packages( 'mvtnorm' )             # Do this only once
library( mvtnorm )                        # Do this each new R-session
sample <- rmvnorm( 100, sigma=diag(1,2) ) # Now it works
plot( sample )
```

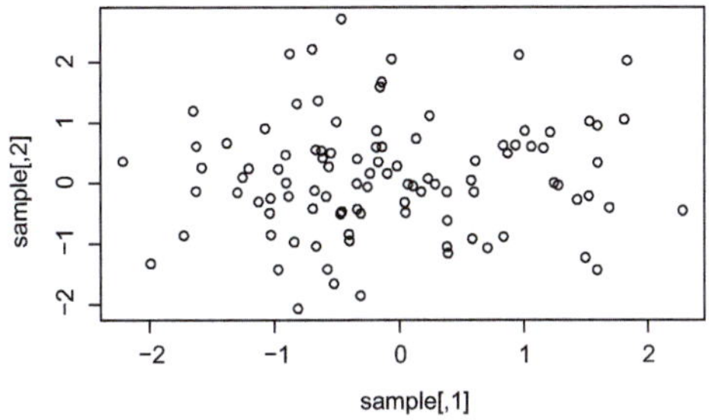

Appendix E
Bayesian Software

The code examples in this book were for the most part written in basic R, and they were often preceded by the relevant mathematical equations. Only on a few occasions were software packages with pre-defined functions used, and then only after examples in basic R had been given first. The idea was to avoid the black-box syndrome where you learn how to produce numbers but are not sure what you are doing, and cannot improve the algorithm. Despite the reliance on basic R, all code-chunks in this book were very short which confirmed the underlying simplicity of Bayesian methods. But you know all that now and may want to start using software that makes the implementation, execution, and analysis of Bayesian methods just a little bit easier and faster. And free software exists that makes it easy to encode priors and likelihood functions and run an MCMC. These include, in historical order, the packages:

- BUGS (and more recently WinBUGS and OpenBUGS) (https://www.mrc-bsu.cam.ac.uk/software),
- JAGS (https://sourceforge.net/projects/mcmc-jags),
- Stan (https://mc-stan.org),
- Nimble (https://r-nimble.org).

For a longer overview of these packages and comparable software for other languages than R, mainly Python, see Štrumbelj et al. (2024). Unfortunately, these packages are currently mainly suited for hierarchical models with well-defined conditional probability distributions and not for complex PBMs. However, in recent years, two impressive pieces of software have been developed that facilitate the Bayesian calibration of PBMs using MCMC:

- BayesianTools (Hartig et al., 2017) can run different MCMC algorithms, and it has been incorporated in ...
- PEcAn (Dietze et al., 2013) which allows the user to choose from a suite of different PBMs and driver conditions.

M. van Oijen, *Bayesian Compendium*,
https://doi.org/10.1007/978-3-031-66085-6

But if you have any specific statistical model or computational tool in mind, and you wonder whether any R-packages exist for that, your first port of call should be the CRAN repository of R-software, and in particular its 'Task View for Bayesian Inference': https://cran.r-project.org/web/views/Bayesian.html. There is a wealth of material there, including a link to `geoR` which we used in Chaps. 15 and 23.

Appendix F
Solutions to Exercises

Many exercises ask you to study or change the R-code used for this book. You can download the code from the author's GitHub repository: https://doi.org/10.5281/zenodo.11092851.

Chapter 1

1. Uncertainty propagation. The predictive uncertainties of the four models can be represented as follows:

- **a.** $z \sim N[3, 2]$.
- **b.** $z \sim N[-1, 2]$.
- **c.** `z <- rnorm(1e4,1,1) * rnorm(1e4,2,1)`.
- **d.** `z <- exp( -(rnorm(1e4,1,1) + rnorm(1e4,2,1))**2 )`.

Probability problems usually have no analytical solutions, so **a** and **b** are exceptions. Sampling is almost always possible, but make your code more readable than the one-liners we gave for **c** and **d**! Here is sampling code for all four models that also produces histograms.

```
fa <- function(x,b) {x + b} ; fb <- function(x,b) {x - b}
fc <- function(x,b) {x * b} ; fd <- function(x,b) {exp(-(x + b)**2)}
n  <- 1e4 ; x <- rnorm(n,1,1) ; b <- rnorm(n,2,1)
par( mfrow=c(1,4) ) # This command prepares 1 row of 4 plots
hist( fa(x,b) ) ; hist( fb(x,b) ) ; hist( fc(x,b) ) ; hist(fd(x,b) )
```


M. van Oijen, *Bayesian Compendium*,
https://doi.org/10.1007/978-3-031-66085-6

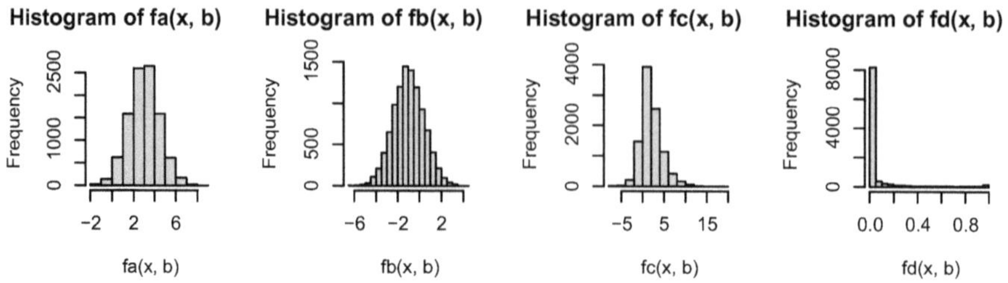

Note that the predictive uncertainty, as measured by the variance, is the same in **a** and **b**, slightly higher in **c** and much smaller in **d**, even though input uncertainty was the same in each case. Uncertainty propagation depends very much on model structure.

Chapter 2

1. **Simpson's Paradox (part 1).** No, higher conditional probabilities of running ($p[run|old] > p[run|young]$) in both subgroups does not imply that the conditional probability is higher when the groups are combined. See the answer to the next question for a counter-example. This is an example of *Simpson's Paradox* (Bernardo & Smith, 1994; Simpson, 1951).
2. **Simpson's Paradox (part 2).**

   - **a.** Smokers: $p[run|young] = 0/2$; $p[run|old] = 1/8$.
   - **b.** Non-smokers: $p[run|young] = 3/8$; $p[run|old] = 1/2$.
   - **c.** Overall: $p[run|young] = 3/10$; $p[run|old] = 2/10$.

   The paradoxical result that being old gives a higher probability of running in smokers (**a**) and in non-smokers (**b**), but not in the combined group (**c**), comes about because smoking and age are not independent properties in this population. By controlling for smoking, we negated the effect of age on running.
3. **Derivation of Bayes' Theorem.** Bayes' Theorem follows directly from the Product Rule, which tells us that $p[A]\,p[B|A] = p[B]\,p[A|B]$. Just divide both sides by $p[A]$.
4. **Medical problem, different prior odds.** The prior odds are now 5/95, but the likelihood ratio is still 99. So your posterior odds of infection are $5/95 * 99 \approx 5.21$. That implies a posterior probability of $5.21/(1 + 5.21) \approx 0.84$. That is much higher than in the original example: the prior is important!
5. **Medical problem, different likelihood ratio.** The prior odds are still 1/99, but with a negative test result the likelihood ratio is also just 1/99, so your posterior odds are $(1/99)^2$. That equals a posterior probability of infection of just over 0.0001.
6. **Inference when data becomes overwhelming.** Your assistant's relative probability distribution for wife:son:butler would go from 60:20:20 via 75:25:0 to

60:40:0 percent (or equivalently from 3:1:1 via 3:1:0 to 3:2:0). And yes, with continuing data accumulation, your distributions will keep converging.

Chapter 3

1. **Family matters (prior).**
   - **a.** You assigned a Jeffreys prior, i.e. $p[N] \propto 1/N$. So the odds on $N \in 1..6$ are $1 : \frac{1}{2} : \frac{1}{3} : \frac{1}{4} : \frac{1}{5} : \frac{1}{6}$ which is equivalent to $60 : 30 : 20 : 15 : 12 : 10$.
   - **b.** $p[\text{Only child}] = p[N = 1] = 1/\sum \frac{1}{n}$, where the summation is over $n = 1..6$. This equates to $60/(60 + 30 + 20 + 15 + 12 + 10) = 60/147 \approx 0.41$.
   - **c.** $p[\text{Eldest}] = \sum (p[N = n] * p[\text{Eldest}|N = n]) = \sum \left(p[N = n] * \frac{1}{n}\right)$. This simplifies to $\sum \frac{1}{n^2} / \sum \frac{1}{n} \approx 0.61$.

   Short R-code for **b** and **c** is as follows:

```
p.b <- 1 / sum( 1/(1:6) )
p.c <- sum( 1/(1:6)^2 ) / sum( 1/(1:6) )
```

Chapter 4

1. **Family matters (likelihood).** The likelihood function is the probability of the data for any value of the parameter of interest. So here it is the probability that Tom happens to be the n-th child if he is one out of N children. Clearly, that probability is zero if $N < n$. For $N \geq n$, the probability is $1/N$.
2. **Correlated data errors.** With very strong correlations (ρ close to 1), you expect the estimate of the intercept to be worse, but that of the slope to be better. That is because all data will then be shifted upward or downward from the true line by a near-constant amount. In other words, the unknown additive measurement error is then systematic rather than stochastic (see Eq. (4.2)).

Chapter 5

1. **Family matters (posterior).** *This question completes the family matters exercises that we studied in Chaps. 3 (prior) and 4 (likelihood).* Because Tom is the n-th eldest, the likelihood (and therefore also the posterior probability) for $N < n$ is zero. For $N \geq n$, both prior and likelihood are proportional to $1/N$, so the posterior is proportional to $1/N^2$. That fully defines the posterior distribution. You can write it formally as $p[N|y] = \frac{1}{N^2} / \sum_{j=n}^{6} \frac{1}{j^2}$ for $N \in n..6$.

2. **Inference from the analytical formula.** If $n = 0$ then the posterior is the same as the prior: $\mu_1 = \mu_0$ and $\sigma_1 = \sigma_0$. If $n \to \infty$ then the prior becomes irrelevant, and the uncertainty disappears: $\mu_1 = \bar{y}$ and $\sigma_1 \to 0$.
3. **Explore the A-R algorithm.** This requires self-study by modifying the downloadable code for this chapter.

Chapter 6

1. **Metropolis for sampling from the posterior.** The R-function for Metropolis that we defined in this chapter (`Metropolis(p, b0, SProp, ni)`) can be called with the product of prior and likelihood as its first argument, because all we need to provide that function with is a quantity that is proportional to our desired posterior distribution, and Bayes' Theorem tells us that the product of prior and likelihood fits the bill.

Chapter 7

1. **Identifiability of parameters.** The posterior parameter uncertainties will be much larger for the model with redundancy. And even if the number of data points would go to infinity, the uncertainty in the two slope parameters of the second model would remain large. But the uncertainty about their *sum* would approach zero. So both models (2- and 3-parameter versions) produce equally reliable output predictions after calibration. We conclude that parameter identifiability is generally a non-issue, unless for some reason you are more interested in the values of your model's parameters than in the model's predictions. Of course, there are other good reasons for removing redundant parameters, primarily that of keeping the model simple.
2. **Convergence and posterior.** The MCMCs will only behave identically if parameter settings for the prior and proposal are carefully chosen. For the first parameter of both models (their intercepts $\beta_1^{(1)}$ resp. $\beta_1^{(2)}$), you can use the same settings for prior and proposal. But to ensure that the MCMCs explore parameter space equally efficiently for the *slope parameter* of the first model ($\beta_2^{(1)}$) as for the *sum of the two slope parameters* ($\beta_2^{(2)} + \beta_3^{(2)}$) of the second model, the settings need to be different. The *sum* of the prior means, of the prior variances, and of the proposal variances for $\beta_2^{(2)}$ and $\beta_3^{(2)}$ should equal the corresponding values for $\beta_2^{(1)}$. Only then will the MCMCs produce the same posterior distribution for $\beta_2^{(1)}$ as for $\beta_2^{(2)} + \beta_3^{(2)}$, and thus also the same posterior predictive distribution for model output.
3. **Which MCMC produced the following posterior samples?** Panel A = MCMC 1; B = 6; C = 2 ; D = 5; E = 3; F = 4.

Chapter 8

1. **Parameter correlations.** Just add '`printcor(bChain)`' to the code. That command derives from the MCMC chain the correlation matrix with all ten posterior parameter correlation coefficients ρ_{ij}. You will then see that there are no instances of abs(ρ_{ij}) exceeding 0.7. That suggests that the model is not overparameterised.

Chapter 11

1. **Model precision vs. accuracy.** You find the answer by estimating the integrated likelihoods for both models, which you could do with the following lines of code.

```
IL1 <- mean( dnorm( rnorm( 1e4, 10- 8, 1 ), 0, 3 ) )
IL2 <- mean( dnorm( rnorm( 1e4, 10-10, 4 ), 0, 3 ) )
pM1 <- IL1 / (IL1+IL2) ; pM2 <- 1 - pM1
```

In this case, the precise model M1 has the highest posterior probability (about 56%) because it produces fewer very bad predictions than the more accurate model M2.

Chapter 14

1. **Prior uncertainty.** The posterior estimates for β become smaller, moving slightly in the direction of the prior mean of (0,0).
2. **LS72-equations.** If you set $\Sigma_y^{-1} = 0$, the data are not informative, and the equations reduce to $\Sigma_{\beta|y} = \Sigma_\beta$ and $\mu_{\beta|y} = \mu_\beta$. In other words, the posterior then equals the prior.
3. **Kalman Filtering (KF) without intercept uncertainty.** By setting $\Sigma_\beta[1, 1]$ to zero, you have effectively removed all uncertainty about the intercept: you fixed its value at the value that we specified in the chapter: zero. That changed the problem from estimating a two-parameter straight line to fitting a one-parameter line through the origin. The posterior uncertainty for the intercept remains zero (as it was in your modified prior), and the uncertainty for the slope is reduced because we now only needed to fit one parameter. Whereas very high or low values for the slope could previously be compensated for by low resp. high values of the intercept, that is no longer possible when we keep the intercept fixed.

Chapter 15

1. **GP with small correlation length.** If you set $\phi = 0$, you will have carried out the linear regression with `geoR` that was announced in Chap. 14. Your results will also resemble those of the last panel of Fig. 15.1, which by neglecting spatial correlation is effectively linear regression.

Chapter 16

1. **Bayes' Theorem as a DAG.** $[\theta] \rightarrow [y]$ is the DAG for us, because $p[y|\theta]$ is easier to specify than $p[\theta|y]$. Of course, Bayes' Theorem will help us to track back from y to θ when measurements of y become available.
2. **Covariance matrices (after MacKay).** Convert the covariance matrices M1 .. M4 to precision matrices using R's `solve()` function, and then apply the `VcondR()` function, defined in the chapter, to see which R-matrices have none-zero entries in the required positions [2,1], [3,1], [3,2]. In short, type `VcondR(solve(M1))` etc. The answer to the question is: M3 and M4.
3. **Simplest hierarchical model.**

   - **a.** $p[y = 1|\xi = 1] = \sum_\theta (p[y = 1|\theta]\, p[\theta|\xi = 1])$. The summation over θ is to distinguish the two possible values of the intermediate variable 'smoking'. When we write this out and plug in the numbers, we find that $p[y = 1|\xi = 1] = 0.4 \times 0.2 + 0.1 \times 0.8 = 0.16$.
   - **b.** $p[y = 1|\xi = 0] = 0.4 \times 0.8 + 0.1 \times 0.2 = 0.34$.
   - **c.** Bayes' Theorem tells us that $p[\xi = 1|y = 1] = p[\xi = 1] \times p[y = 1|\xi = 1]/p[y = 1]$ and we have done most of the work for that already in questions **a** and **b**. The answer is $p[\xi = 1|y = 1] = 0.5\times0.16/(0.5\times0.16+0.5\times0.34) = 0.32$. So the information that someone is a runner decreases their probability of being old from 0.5 to 0.32.
   - **d.** You can generate one population of 20 people from this graphical model with the following short code:

```
Age   <- rbinom( 20, 1, 0.5             )
Smoke <- rbinom( 20, 1, 0.2+0.6*Age     )
Run   <- rbinom( 20, 1, 0.4-0.3*Smoke )
```

where `Age` $= \xi$, `Smoke` $= \theta$, `Run` $= y$. if you repeat this 10^4 times (maybe store the results in a list), you should find that about 6% of populations show Simpson's Paradox.

Chapter 17

1. **Taking the H out of BHM.** Yes, the BHM will collapse to a non-hierarchical model if we make the priors for hyperparameter-variances extremely narrow.
2. **Simplest hierarchical model.** See the last exercise for Chap. 16. Graphical modelling is a particularly useful tool for hierarchical models.

Chapter 18

1. **Derivation of V.** We have $V = R \,/\, p[H] = (E[z|\neg H] - E[z]) \,/\, p[H]$. We expand the last numerator term as $E[z] = p[H]E[z|H] + (1 - p[H])E[z|\neg H]$. That gives us $V = (E[z|\neg H] - p[H]E[z|H] - (1 - p[H])E[z|\neg H]) \,/\, p[H]$, which simplifies to $V = E[z|\neg H] - E[z|H]$.
2. **Approximate PRA for bivariate Gaussian** $p[x, z]$. Substitute the conditional expectations of Eq. (18.9) into Eq. (18.2).
3. **Uncertainty as a function of number of data.** Equations (18.11), (18.12), and (18.13) imply that the uncertainties expressed as variances will be close to halved when sample size is doubled. You can check this numerically using the R-function defined in the chapter: standard deviations generated by `PRA.UQ(rep(x, 2), rep(z, 2))` are a factor $\sqrt{2}$ smaller than those from `PRA.UQ(x, z)`.

Chapter 19

1. **Linear BDT.** The GBN depicted in Fig. 19.2 is a linear model, so $E[u]$ varies linearly with irrigation—we can ignore the variances. We can find the optimum irrigation level by reading just three numbers from the figure. The average cost per mm of irrigation is shown to be 0.5. One mm of irrigation is expected to produce 0.01 unit of extra growth at a benefit per unit growth of 30, so the average benefit of one mm of irrigation is only 0.3. Therefore irrigation cost always exceeds benefit, and the optimum irrigation level is zero.
2. **Life expectancy.** The expected benefit of taking the medicine is $B = p[H](E[z|H, M] - E[z|H, \neg M]) = 50 - x/2 - x^2/200$. Its expected cost is $C = p[\neg H](E[z|\neg H, \neg M] - E[z|\neg H, M]) = x^2/200$. Solve the quadratic equation to find that benefit is expected to equal cost at $x = 50$. So you should only take the medicine if you are below 50 years of age.

Chapter 20

1. **Explore the R-packages.** You may want to begin with the R-commands `??rstanarm` and `??brms` to get information on the packages from within R.

Chapter 21

1. **Classifying disease.** There is no straightforward answer to this. The value of Naive Bayes here depends on how independent the test results are. If you asked the same lab to redo the analysis on your original blood sample, then the results will likely be the same and Naive Bayes should not be used. But if you provided a new sample to another independent laboratory then you may consider the results independent and use Naive Bayes.
2. **The neural network of Fig. 21.2.** Here $n_p = n_h(n_x + 1) + n_y(n_h + 1) = 3 \times 3 + 2 \times 4 = 17$. For this small network, you could also see this directly from the figure: 12 arrow-weights plus 5 neuron- and output-biases.
3. **1-1-1 neural networks.** Choose identity activation functions, biases equal to zero, and weights such that their product $\texttt{w1} * \texttt{w2} = 2$. For example: `fNN.1n1(x, 0, 0, 1, 2, identity)` or `fNN.1n1(x, 0, 0, 2, 1, identity)`.

Chapter 22

1. **Smoothness of a function represented by a GP**. Choose a larger value for `phi`, e.g. 100 or 1000.
2. **Observational operator in KF. H** = (0, 0.8, 0).
3. **Bivariate vs. univariate KF.** To make the bivariate KF equivalent to the univariate one, double the predictive uncertainty variances, i.e. set `Sf <- diag(2,2)`.

Chapter 23

1. **Scaling error.** The second derivative of x^3 is $6x$. So $\widehat{\Delta} = -\frac{1}{2}\text{Var}[x] f^{(2)}(E[x]) = -\frac{1}{2} * 10 * 30 = -150$. So our estimate of 125 has an error equal to -150, and the corrected answer therefore is 275. This answer is exact (for the proof, see Van Oijen et al., 2017), but you can check it numerically by sampling in R:

```
x <- rnorm(1e5, 5, sqrt(10)) ; z <- x^3 ; meanz <- mean(z)
```

Chapter 24

1. **Explore adaptive sampling.** This requires self-study by modifying the downloadable code for this chapter.

References

Anderson, E. (1936). The species problem in Iris. *Annals of the Missouri Botanical Garden, 23,* 457–509. https://doi.org/10.2307/2394164

Andrianakis, I., & Challenor, P. G. (2012). The effect of the nugget on Gaussian process emulators of computer models. *Computational Statistics & Data Analysis, 56,* 4215–4228. https://doi.org/10.1016/j.csda.2012.04.020

Andrianakis, I., Vernon, I. R., McCreesh, N., McKinley, T. J., Oakley, J. E., Nsubuga, R. N., et al. (2015). Bayesian history matching of complex infectious disease models using emulation: a tutorial and a case study on HIV in Uganda. *PLoS Computational Biology, 11,* e1003968. https://doi.org/10.1371/journal.pcbi.1003968

Andrieu, C., & Thoms, J. (2008). A tutorial on adaptive MCMC. *Statistics and Computing, 18,* 343–373. https://doi.org/10.1007/s11222-008-9110-y

Bailer-Jones, C. A. L. (2017). *Practical Bayesian inference: A primer for physical scientists.* Cambridge University Press.

Bakka, H., Rue, H., Fuglstad, G.-A., Riebler, A., Bolin, D., Krainski, E., Simpson, D., & Lindgren, F. (2018). Spatial modelling with R-INLA: A review. arXiv:1802.06350 [stat].

Barber, D., & Bishop, C. M. (1998). Ensemble learning in Bayesian neural networks. In *Generalization in neural networks and machine learning* (pp. 215–238). Springer.

Bayes, T., (1763). An essay towards solving a problem in the doctrine of chances. *Philosophical Transactions, 53,* 370–418. https://doi.org/10.1098/rstl.1763.0053

Bengio, Y., Ducharme, R., Vincent, P., & Jauvin, C. (2003). A neural probabilistic language model. *Journal of Machine Learning Research, 3,* 1137–1155.

Berger, J. O. (1985). *Statistical decision theory and bayesian analysis* (2nd ed.). Springer Series in Statistics. Springer.

Bernardo, J. M., & Smith, A. F. M. (1994). *Bayesian theory.* Wiley.

Betancourt, M. (2018). A Conceptual Introduction to Hamiltonian Monte Carlo. arXiv:1701.02434 [stat].

Bishop, C. M. (2006). *Pattern recognition and machine learning, information science and statistics.* Springer.

Blangiardo, M., Cameletti, M., Baio, G., & Rue, H. (2013). Spatial and spatio-temporal models with R-INLA. *Spatial and Spatio-temporal Epidemiology, 4,* 33–49. https://doi.org/10.1016/j.sste.2012.12.001

Blei, D. M., Kucukelbir, A., & McAuliffe, J. D. (2017). Variational Inference: A review for statisticians. *Journal of the American Statistical Association, 112,* 859–877. https://doi.org/10.1080/01621459.2017.1285773

Bolker, B. M. (2008). *Ecological models and data in R.* Princeton University Press.

M. van Oijen, *Bayesian Compendium*,
https://doi.org/10.1007/978-3-031-66085-6

Bon, J. J., Bretherton, A., Buchhorn, K., Cramb, S., Drovandi, C., Hassan, C., et al. (2023). Being Bayesian in the 2020s: Opportunities and challenges in the practice of modern applied Bayesian statistics. *Philosophical Transactions of the Royal Society A: Mathematical, Physical and Engineering Sciences, 381,* 20220156. https://doi.org/10.1098/rsta.2022.0156

Borgonovo, E., Cappelli, V., Maccheroni, F., & Marinacci, M. (2018). Risk analysis and decision theory: A bridge. *European Journal of Operational Research, 264,* 280–293. https://doi.org/10.1016/j.ejor.2017.06.059.

Bradley, W., Kim, J., Kilwein, Z., Blakely, L., Eydenberg, M., Jalvin, J., et al. (2022). Perspectives on the integration between first-principles and data-driven modeling. *Computers & Chemical Engineering, 166,* 107898. https://doi.org/10.1016/j.compchemeng.2022.107898

Briganti, G., Scutari, M., & McNally, R. J. (2023). A tutorial on bayesian networks for psychopathology researchers. *Psychological Methods, 28,* 947–961. https://doi.org/10.1037/met0000479

Brynjarsdóttir, J., & O'Hagan, A. (2014). Learning about physical parameters: The importance of model discrepancy. *Inverse Problems, 30,* 114007. https://doi.org/10.1088/0266-5611/30/11/114007

Burnham, K. P., & Anderson, D. R. (2002). *Model selection and multimodel inference: A practical information-theoretic approach.* Springer.

Cameron, D., Hartig, F., Minnuno, F., Oberpriller, J., Reineking, B., Van Oijen, M., & Dietze, M. (2022). Issues in calibrating models with multiple unbalanced constraints: The significance of systematic model and data errors. *Methods in Ecology and Evolution,* 2041–210X.14002. https://doi.org/10.1111/2041-210X.14002

Cameron, D. R., Van Oijen, M., Werner, C., Butterbach-Bahl, K., Grote, R., Haas, E., et al. (2013). Environmental change impacts on the C- and N-cycle of European forests: A model comparison study. *Biogeosciences, 10,* 1751–1773. https://doi.org/10.5194/bg-10-1751-2013

Carbajal, J. P., Leitão, J. P., Albert, C., & Rieckermann, J. (2017). Appraisal of data-driven and mechanistic emulators of nonlinear simulators: The case of hydrodynamic urban drainage models. *Environmental Modelling & Software, 92,* 17–27. https://doi.org/10.1016/j.envsoft.2017.02.006

Castillo, E., Menéndez, J. M., & Sánchez-Cambronero, S. (2008). Predicting traffic flow using Bayesian networks. *Transportation Research Part B: Methodological, 42,* 482–509. https://doi.org/10.1016/j.trb.2007.10.003

Chamberlin, T. C. (1890). The method of multiple working hypotheses. *Science, 15,* 92–96.

Chandler, R. E. (2013). Exploiting strength, discounting weakness: Combining information from multiple climate simulators. *Philosophical Transactions of the Royal Society A: Mathematical, Physical and Engineering Sciences, 371,* 20120388–20120388. https://doi.org/10.1098/rsta.2012.0388

Chipman, H. A., George, E. I., & McCulloch, R. E. (2010). BART: Bayesian additive regression trees. *The Annals of Applied Statistics, 4,* 266–298. https://doi.org/10.1214/09-AOAS285

Clark, J. S., Iverson, L., Woodall, C. W., Allen, C. D., Bell, D. M., Bragg, D. C., et al. (2016). The impacts of increasing drought on forest dynamics, structure, and biodiversity in the United States. *Global Change Biology, 22,* 2329–2352. https://doi.org/10.1111/gcb.13160

Conti, S., & O'Hagan, A. (2010). Bayesian emulation of complex multi-output and dynamic computer models. *Journal of Statistical Planning and Inference, 140,* 640–651. https://doi.org/10.1016/j.jspi.2009.08.006

Cressie, N., Calder, C. A., Clark, J. S., Hoef, J. M. V., & Wikle, C. K. (2009). Accounting for uncertainty in ecological analysis: The strengths and limitations of hierarchical statistical modeling. *Ecological Applications, 19,* 553–570.

Cressie, N., & Wikle, C. K. (2011). *Statistics for spatio-temporal data.* Wiley.

Deng, L. (2012). The MNIST database of handwritten digit images for machine learning research [Best of the Web]. *IEEE Signal Processing Magazine, 29,* 141–142. https://doi.org/10.1109/MSP.2012.2211477

Dey, D., Datta, A., & Banerjee, S. (2022). Graphical Gaussian process models for highly multivariate spatial data. *Biometrika, 109,* 993–1014. https://doi.org/10.1093/biomet/asab061

Dietze, M. C., Lebauer, D. S., & Kooper, R. (2013). On improving the communication between models and data. *Plant, Cell & Environment, 36,* 1575–1585. https://doi.org/10.1111/pce.12043

Dietze, M. C., Wolosin, M. S., & Clark, J. S. (2008). Capturing diversity and interspecific variability in allometries: A hierarchical approach. *Forest Ecology and Management, 256,* 1939–1948. https://doi.org/10.1016/j.foreco.2008.07.034

Diggle, P., & Ribeiro, P. J. (2007). *Model-based geostatistics*. Springer.

Diggle, P. J., Ribeiro Jr, P. J., & Christensen, O. F. (2003). An introduction to model-based geostatistics. In *Spatial statistics and computational methods* (pp. 43–86). Springer.

Drignei, D. (2017). An estimation algorithm for fast kriging surrogates of computer models with unstructured multiple outputs. *Computer Methods in Applied Mechanics and Engineering, 321,* 35–45. https://doi.org/10.1016/j.cma.2017.04.001

DuMouchel, W., & Normand, S.-L. (2000). Computer-modeling and graphical strategies for meta-analysis. In *Meta-analysis in medicine and health policy* (pp. 127–178).

Dunstan, D. J., Crowne, J., & Drew, A. J. (2022). Easy computation of the Bayes factor to fully quantify Occam's razor in least-squares fitting and to guide actions. *Scientific Reports, 12,* 993. https://doi.org/10.1038/s41598-021-04694-7

Fer, I., Shiklomanov, A. N., Novick, K. A., Gough, C. M., Arain, M. A., Chen, J., et al. (2021). Capturing site-to-site variability through hierarchical Bayesian calibration of a process-based dynamic vegetation model (Preprint). *Ecology*. https://doi.org/10.1101/2021.04.28.441243

Fu, Y. H., Campioli, M., Van Oijen, M., Deckmyn, G., & Janssens, I. A. (2012). Bayesian comparison of six different temperature-based budburst models for four temperate tree species. *Ecological Modelling, 230,* 92–100. https://doi.org/10.1016/j.ecolmodel.2012.01.010

Gabry, J., Simpson, D., Vehtari, A., Betancourt, M., & Gelman, A. (2019). Visualization in Bayesian workflow. *Journal of the Royal Statistical Society: Series A (Statistics in Society), 182,* 389–402. https://doi.org/10.1111/rssa.12378

Gelman, A., Carlin, J. B., Stern, H. S., Dunson, D. B., Vehtari, A., & Rubin, D. B. (2013). *Bayesian data analysis* (3rd ed.). Chapman and Hall/CRC.

Gelman, A., & Hill, J. (2006). *Data analysis using regression and multilevel/hierarchical models*. Cambridge University Press.

Gelman, A., Hwang, J., & Vehtari, A. (2014). Understanding predictive information criteria for Bayesian models. *Statistics and Computing, 24,* 997–1016. https://doi.org/10.1007/s11222-013-9416-2

Ghahramani, Z. (2015). Probabilistic machine learning and artificial intelligence. *Nature, 521,* 452–459. https://doi.org/10.1038/nature14541

Gilks, W. R., Richardson, S., & Spiegelhalter, D. (1995). *Markov chain Monte Carlo in practice*. CRC Press.

Goldstein, M. (2015). Bayes linear analysis. In *Wiley StatsRef: Statistics Reference Online* (pp. 1–7). American Cancer Society. https://doi.org/10.1002/9781118445112.stat00225.pub2

Goldstein, M. (2011). External Bayesian analysis for computer simulators. *Bayesian statistics, 9,* 201–228. https://doi.org/10.1093/acprof:oso/9780199694587.003.0007

Goldstein, M., & Rougier, J. (2009). Reified Bayesian modelling and inference for physical systems. *Journal of Statistical Planning and Inference, 139,* 1221–1239. https://doi.org/10.1016/j.jspi.2008.07.019

Gonzalez-Redin, J., Luque, S., Poggio, L., Smith, R., & Gimona, A. (2016). Spatial Bayesian belief networks as a planning decision tool for mapping ecosystem services trade-offs on forested landscapes. *Environmental Research, The Provision of Ecosystem Services in Response to Global Change, 144,* 15–26. https://doi.org/10.1016/j.envres.2015.11.009

Goudriaan, J., & Monteith, J. L. (1990). A mathematical function for crop growth based on light interception and leaf area expansion. *Annals of Botany, 66,* 695–701.

Green, P. J. (1995). Reversible jump Markov chain Monte Carlo computation and Bayesian model determination. *Biometrika, 82,* 711–732. https://doi.org/10.1093/biomet/82.4.711

Gregory, P. (2005). *Bayesian logical data analysis for the physical sciences: A comparative approach with mathematica® support*. Cambridge University Press. https://doi.org/10.1017/CBO9780511791277

Haario, H., Laine, M., Mira, A., & Saksman, E. (2006). DRAM: Efficient adaptive MCMC. *Statistics and Computing, 16,* 339–354. https://doi.org/10.1007/s11222-006-9438-0

Hahn, U. (2014). The Bayesian boom: Good thing or bad? *Frontiers in Psychology, 5,* 765. https://doi.org/10.3389/fpsyg.2014.00765

Hanson, B., Stall, S., Cutcher-Gershenfeld, J., Vrouwenvelder, K., Wirz, C., Rao, Y. (Douglas)., & Peng, G. (2023). Garbage in, garbage out: Mitigating risks and maximizing benefits of AI in research. *Nature, 623,* 28–31. https://doi.org/10.1038/d41586-023-03316-8

Hartig, F., Dyke, J., Hickler, T., Higgins, S. I., O'Hara, R. B., Scheiter, S., & Huth, A. (2012). Connecting dynamic vegetation models to data - an inverse perspective. *Journal of Biogeography, 39,* 2240–2252. https://doi.org/10.1111/j.1365-2699.2012.02745.x

Hartig, F., Minunno, F., Paul, S., & Cameron, D. (2017). *Package 'BayesianTools'.*

Hauser, T., Keats, A., & Tarasov, L. (2012). Artificial neural network assisted Bayesian calibration of climate models. *Climate Dynamics, 39,* 137–154. https://doi.org/10.1007/s00382-011-1168-0

Hayes, B. (2013). First links in the Markov chain. *American Scientist, 101,* 92–97.

Hickler, T., Rammig, A., & Werner, C. (2015). Modelling CO_2 impacts on forest productivity. *Current Forestry Reports, 1,* 69–80. https://doi.org/10.1007/s40725-015-0014-8

Higdon, D., Gattiker, J., Williams, B., & Rightley, M. (2008). Computer model calibration using high-dimensional output. *Journal of the American Statistical Association, 103,* 570–583. https://doi.org/10.1198/016214507000000888

Hines, W. W., & Montgomery, D. C. (1990). *Probability and statistics in engineering and management science* (3rd ed.). Wiley.

Hjelkrem, A.-G.R., Höglind, M., van Oijen, M., Schellberg, J., Gaiser, T., & Ewert, F. (2017). Sensitivity analysis and Bayesian calibration for testing robustness of the BASGRA model in different environments. *Ecological Modelling, 359,* 80–91. https://doi.org/10.1016/j.ecolmodel.2017.05.015

HM_Government (2020). *National Risk Register.* https://www.gov.uk/government/publications/national-risk-register-2020

Höglind, M., Van Oijen, M., Cameron, D., & Persson, T. (2016). Process-based simulation of growth and overwintering of grassland using the BASGRA model. *Ecological Modelling, 335,* 1–15. https://doi.org/10.1016/j.ecolmodel.2016.04.024

Hojsgaard, S., Edwards, D., & Lauritzen, S. (2012). *Graphical Models with R.*

Huang, H.-C., & Cressie, N. (1996). Spatio-temporal prediction of snow water equivalent using the Kalman filter. *Computational Statistics & Data Analysis, 22,* 159–175. https://doi.org/10.1016/0167-9473(95)00047-X

Huard, D., Évin, G., & Favre, A.-C. (2006). Bayesian copula selection. *Computational Statistics & Data Analysis, 51,* 809–822. https://doi.org/10.1016/j.csda.2005.08.010

Hullman, J., Qiao, X., Correll, M., Kale, A., & Kay, M. (2019). In pursuit of error: A survey of uncertainty visualization evaluation. *IEEE Transactions on Visualization and Computer Graphics, 25,* 903–913. https://doi.org/10.1109/TVCG.2018.2864889

Hyvönen, R., Ågren, G. I., Linder, S., Persson, T., Cotrufo, M. F., Ekblad, A., et al. (2007). The likely impact of elevated [CO_2], nitrogen deposition, increased temperature and management on carbon sequestration in temperate and boreal forest ecosystems: A literature review. *New Phytologist, 173,* 463–480. https://doi.org/10.1111/j.1469-8137.2007.01967.x

James, G., Witten, D., Hastie, T., & Tibshirani, R. (2013). *An introduction to statistical learning, springer texts in statistics.* Springer. https://doi.org/10.1007/978-1-4614-7138-7

Jaynes, E. T. (2003). *Probability theory: The logic of science.* Cambridge University Press.

Johnson, M. O., Galbraith, D., Gloor, M., De Deurwaerder, H., Guimberteau, M., Rammig, A., et al. (2016). Variation in stem mortality rates determines patterns of above-ground biomass in Amazonian forests: Implications for dynamic global vegetation models. *Global Change Biology, 22,* 3996–4013. https://doi.org/10.1111/gcb.13315

Kapelner, A., & Bleich, J. (2016). bartMachine: Machine learning with Bayesian additive regression trees. *Journal of Statistical Software, 70,* 40. https://doi.org/10.18637/jss.v070.i04

Kass, R. E., & Raftery, A. E. (1995). Bayes factors. *Journal of the American Statistical Association, 90,* 773–795.

Kavetski, D., Kuczera, G., & Franks, S. W. (2006). Bayesian analysis of input uncertainty in hydrological modeling: 1. Theory. *Water Resources Research, 42.* https://doi.org/10.1029/2005WR004368

Kennedy, M. C., & O'Hagan, A. (2001). Bayesian calibration of computer models. *Journal of the Royal Statistical Society: Series B (Statistical Methodology), 63,* 425–464.

Kobayashi, K., & Salam, M. U. (2000). Comparing simulated and measured values using mean squared deviation and its components. *Agronomy Journal, 92,* 345–352. https://doi.org/10.2134/agronj2000.922345x

Kuhn, M. (2008). Building predictive models in R using the caret package. *Journal of Statistical Software, 28*, 1–26. https://doi.org/10.18637/jss.v028.i05

Lam, R., Sanchez-Gonzalez, A., Willson, M., Wirnsberger, P., Fortunato, M., Alet, F., et al. (2023). Learning skillful medium-range global weather forecasting. *Science, 382,* 1416–1421. https://doi.org/10.1126/science.adi2336

Leeds, W. B., Wikle, C. K., Fiechter, J., Brown, J., & Milliff, R. F. (2013). Modeling 3-D spatio-temporal biogeochemical processes with a forest of 1-D statistical emulators: Modeling 3-D processes with a forest of 1-D emulators. *Environmetrics, 24,* 1–12. https://doi.org/10.1002/env.2187

Levy, P., Clement, R., Cowan, N., Keane, B., Myrgiotis, V., van Oijen, M., Smallman, T.L., Toet, S., & Williams, M. (2022). Challenges in scaling up greenhouse gas fluxes: Experience from the UK greenhouse gas emissions and feedbacks program. *Journal of Geophysical Research: Biogeosciences, 127*, e2021JG006743. https://doi.org/10.1029/2021JG006743

Levy, P. E., Cowan, N., van Oijen, M., Famulari, D., Drewer, J., & Skiba, U. (2017). Estimation of cumulative fluxes of nitrous oxide: Uncertainty in temporal upscaling and emission factors: Estimation of cumulative fluxes of nitrous oxide. *European Journal of Soil Science, 68,* 400–411. https://doi.org/10.1111/ejss.12432

Levy, P. E., Wendler, R., Van Oijen, M., Cannell, M. G., & Millard, P. (2005). The effect of nitrogen enrichment on the carbon sink in coniferous forests: Uncertainty and sensitivity analyses of three ecosystem models. *Water, Air, & Soil Pollution: Focus, 4,* 67–74.

Li, T., Zhang, G., Do, Q. D., Yue, X., & Chen, W. (2024). *Long-context LLMs Struggle with Long In-context Learning*. https://doi.org/10.48550/arXiv.2404.02060

Lindley, D. V. (1991). *Making decisions* (2nd ed.). Wiley.

Lindley, D. V., & Smith, A. F. M. (1972). Bayes estimates for the linear model. *Journal of the Royal Statistical Society Series B: Statistical Methodology, 34,* 1–41.

Lintusaari, J., Gutmann, M. U., Dutta, R., Kaski, S., & Corander, J. (2017). Fundamentals and recent developments in approximate Bayesian computation. *Systematic Biology, 66,* e66–e82. https://doi.org/10.1093/sysbio/syw077

Liu, Y., Attema, J., & Hazeleger, W. (2020). Exploring Bayesian deep learning for weather forecasting with the Lorenz 84 system. In: *ECMWF-ESA Workshop on Machine Learning for Earth System Observation and Prediction.* https://doi.org/10.5281/zenodo.4146850

Loeppky, J. L., Sacks, J., & Welch, W. J. (2009). Choosing the sample size of a computer experiment: A practical guide. *Technometrics, 51,* 366–376. https://doi.org/10.1198/TECH.2009.08040

MacKay, D. J. C. (1992a). Bayesian interpolation. *Neural Computation, 4,* 415–447.

MacKay, D. J. C. (1992b). A practical Bayesian framework for backpropagation networks. *Neural Computation, 4,* 448–472. https://doi.org/10.1162/neco.1992.4.3.448

MacKay, D. J. C. (1995). *Bayesian methods for neural networks: Theory and applications*.

MacKay, D. J. C. (2003). *Information theory, inference, and learning algorithms*. Cambridge University Press.

Mäkelä, A., del Río, M., Hynynen, J., Hawkins, M. J., Reyer, C., Soares, P., et al. (2012). Using stand-scale forest models for estimating indicators of sustainable forest management. *Forest Ecology and Management, 285,* 164–178. https://doi.org/10.1016/j.foreco.2012.07.041

Mann, S. F., Pain, R., & Kirchhoff, M. D. (2022). Free energy: A user's guide. *Biology & Philosophy, 37,* 33. https://doi.org/10.1007/s10539-022-09864-z

Margossian, C.C., & Gelman, A. (2024). For how many iterations should we run Markov chain Monte Carlo?

Margossian, C., Vehtari, A., Simpson, D., & Agrawal, R. (2020). Hamiltonian Monte Carlo using an adjoint-differentiated Laplace approximation: Bayesian inference for latent Gaussian models and beyond. In *Advances in neural information processing systems* (pp. 9086–9097). Curran Associates.

Martin, G. M., Frazier, D. T., & Robert, C. P. (2023). Approximating Bayes in the 21st century. *Statistical Science , 39,* 20–45. https://doi.org/10.1214/22-STS875

Mattei, P.-A. (2019). *A Parsimonious Tour of Bayesian Model Uncertainty*. arXiv: Methodology.

McCulloch, W. S., & Pitts, W. (1943). A logical calculus of the ideas immanent in nervous activity. *The Bulletin of Mathematical Biophysics, 5,* 115–133. https://doi.org/10.1007/BF02478259

McElreath, R. (2016). *Statistical rethinking: A Bayesian course with examples in R and stan*. CRC Press.

Medlyn, B. E., Zaehle, S., De Kauwe, M. G., Walker, A. P., Dietze, M. C., Hanson, P. J., et al. (2015). Using ecosystem experiments to improve vegetation models. *Nature Climate Change, 5,* 528–534. https://doi.org/10.1038/nclimate2621

Mee, R. W., & Owen, D. B. (1983). A simple approximation for bivariate normal probabilities. *Journal of Quality Technology, 15,* 72–75. https://doi.org/10.1080/00224065.1983.11978848

Metropolis, N., Rosenbluth, A. W., Rosenbluth, M. N., Teller, A. H., & Teller, E. (1953). Equation of state calculations by fast computing machines. *The Journal of Chemical Physics, 21,* 1087–1092. https://doi.org/10.1063/1.1699114

Mikkola, P., Martin, O. A., Chandramouli, S., Hartmann, M., Pla, O. A., Thomas, O., Pesonen, H., Corander, J., Vehtari, A., Kaski, S., Bürkner, P.-C., & Klami, A. (2023). Prior knowledge elicitation: The past, present, and future. *Bayesian Analysis, 1,* 1–33. https://doi.org/10.1214/23-BA1381

Milne, A. E., Glendining, M. J., Lark, R. M., Perryman, S. A. M., Gordon, T., & Whitmore, A. P. (2015). Communicating the uncertainty in estimated greenhouse gas emissions from agriculture. *Journal of Environmental Management, 160,* 139–153. https://doi.org/10.1016/j.jenvman.2015.05.034

Minunno, F., van Oijen, M., Cameron, D. R., Cerasoli, S., Pereira, J. S., & Tomé, M. (2013a). Using a Bayesian framework and global sensitivity analysis to identify strengths and weaknesses of two process-based models differing in representation of autotrophic respiration. *Environmental Modelling & Software, 42,* 99–115. https://doi.org/10.1016/j.envsoft.2012.12.010

Minunno, F., Van Oijen, M., Cameron, D. R., & Pereira, J. S. (2013b). Selecting parameters for Bayesian calibration of a process-based model: A methodology based on canonical correlation analysis. *SIAM/ASA Journal on Uncertainty Quantification, 1,* 370–385. https://doi.org/10.1137/120891344

Molnar, C. (2023). *Interpretable machine learning*.

Murphy, K. P. (2022). *Probabilistic machine learning: An introduction*. Adaptive computation and machine learning series. The MIT Press.

Murphy, K. P. (2023). *Probabilistic machine learning: Advanced topics*. Adaptive Computation and Machine Learning Series. The MIT Pres.

Nature Editorial. (2023). Stop talking about tomorrow's AI doomsday when AI poses risks today. *Nature, 618,* 885–886. https://doi.org/10.1038/d41586-023-02094-7

Naveed, H., Khan, A. U., Qiu, S., Saqib, M., Anwar, S., Usman, M., Akhtar, N., Barnes, N., & Mian, A. (2024). A Comprehensive Overview of Large Language Models.

Neal, R. M. (1992). Bayesian learning via stochastic dynamics. In *Advances in neural information processing systems* (pp. 475–482). Morgan-Kaufmann.

Neal, R. M. (1996). Priors for infinite networks. In R. M. Neal (Ed.) *Bayesian learning for neural networks*. Lecture Notes in Statistics (pp. 29–53). Springer. https://doi.org/10.1007/978-1-4612-0745-0_2

Nott, D. J., Fan, Y., Marshall, L., & Sisson, S. A. (2014). Approximate Bayesian computation and Bayes' linear analysis: Toward high-dimensional ABC. *Journal of Computational and Graphical Statistics, 23,* 65–86. https://doi.org/10.1080/10618600.2012.751874

O'Hagan, A. (2006). Bayesian analysis of computer code outputs: A tutorial. *Reliability Engineering & System Safety, 91,* 1290–1300. https://doi.org/10.1016/j.ress.2005.11.025

O'Hagan, A. (2012). Probabilistic uncertainty specification: Overview, elaboration techniques and their application to a mechanistic model of carbon flux. *Environmental Modelling & Software, 36,* 35–48. https://doi.org/10.1016/j.envsoft.2011.03.003

O'Neil, C. (2016). *Weapons of math destruction: How big data increases inequality and threatens democracy*. Penguin.

Ogle, K. (2009). Hierarchical Bayesian statistics: Merging experimental and modeling approaches in ecology. *Ecological Applications, 19,* 577–581. https://doi.org/10.1890/08-0560.1

Ogle, K., & Barber, J. J. (2008). Bayesian data-model integration in plant physiological and ecosystem ecology. In U. Lüttge, W. Beyschlag, & J. Murata (Eds.), *Progress in botany* (pp. 281–311). Springer.

Ogle, K., Pathikonda, S., Sartor, K., Lichstein, J. W., Osnas, J. L. D., & Pacala, S. W. (2014). A model-based meta-analysis for estimating species-specific wood density and identifying potential sources of variation. *Journal of Ecology, 102,* 194–208. https://doi.org/10.1111/1365-2745.12178

Opitz, T. (2017). Latent Gaussian modeling and INLA: A review with focus on space-time applications. *Journal de la Société Française de Statistique, 158,* 62–85.

Patenaude, G., Milne, R., Van Oijen, M., Rowland, C. S., & Hill, R. A. (2008). Integrating remote sensing datasets into ecological modelling: A Bayesian approach. *International Journal of Remote Sensing, 29,* 1295–1315. https://doi.org/10.1080/01431160701736414

Piironen, J., & Vehtari, A. (2015). Comparison of Bayesian predictive methods for model selection. arXiv:1503.08650 [cs, stat] .

Polanska, A., Price, M. A., Mancini, A. S., & McEwen, J. D. (2023). Learned harmonic mean estimation of the marginal likelihood with normalizing flows.

Radhakrishnan, A., Beaglehole, D., Pandit, P., & Belkin, M. (2024). Mechanism for feature learning in neural networks and backpropagation-free machine learning models. *Science, 383,* 1461–1467. https://doi.org/10.1126/science.adi5639

Rasmussen, C. E., & Williams, C. K. (2006). *Gaussian processes for machine learning*.

Reichstein, M., Camps-Valls, G., Stevens, B., Jung, M., Denzler, J., Carvalhais, N., & Prabhat, F. (2019). Deep learning and process understanding for data-driven earth system science. *Nature, 566,* 195. https://doi.org/10.1038/s41586-019-0912-1

Reyer, C. (2015). Forest productivity under environmental change–a review of stand-scale modeling studies. *Current Forestry Reports, 1,* 53–68. https://doi.org/10.1007/s40725-015-0009-5

Reyer, C. P. O., Flechsig, M., Lasch-Born, P., & van Oijen, M. (2016). Integrating parameter uncertainty of a process-based model in assessments of climate change effects on forest productivity. *Climatic Change, 137,* 395–409. https://doi.org/10.1007/s10584-016-1694-1

Robert, C. P., & Casella, G. (2010). *Monte Carlo statistical methods*. Springer.

Robert, C. P., Chopin, N., & Rousseau, J. (2009). Harold Jeffreys's theory of probability revisited. *Statistical Science, 24,* 141–172. https://doi.org/10.1214/09-STS284

Rollinson, C. R., Liu, Y., Raiho, A., Moore, D. J. P., McLachlan, J., Bishop, D. A., et al. (2017). Emergent climate and CO_2 sensitivities of net primary productivity in ecosystem models do not agree with empirical data in temperate forests of eastern North America. *Global Change Biology, 23,* 2755–2767. https://doi.org/10.1111/gcb.13626

Rougier, J. (2007). Probabilistic inference for future climate using an ensemble of climate model evaluations. *Climatic Change, 81,* 247–264. https://doi.org/10.1007/s10584-006-9156-9

Rougier, J. (2008). Efficient emulators for multivariate deterministic functions. *Journal of Computational and Graphical Statistics, 17,* 827–843. https://doi.org/10.1198/106186008X384032

Roustant, O., Ginsbourger, D., & Deville, Y. (2012). *DiceKriging, DiceOptim: Two R packages for the analysis of computer experiments by kriging-based metamodeling and optimization.*

Rudin, C., Chen, C., Chen, Z., Huang, H., Semenova, L., & Zhong, C. (2022). Interpretable machine learning: Fundamental principles and 10 grand challenges. *Statistics Surveys, 16,* 1–85. https://doi.org/10.1214/21-SS133

Rule, J. S., Tenenbaum, J. B., & Piantadosi, S. T. (2020). The child as hacker. *Trends in Cognitive Sciences, 24,* 900–915. https://doi.org/10.1016/j.tics.2020.07.005

Rumelhart, D. E., Hinton, G. E., & Williams, R. J. (1986). Learning representations by back-propagating errors. *Nature, 323,* 533–536. https://doi.org/10.1038/323533a0

Russell, B. (1912). *The problems of philosophy.*

Saltelli, A. (Ed.) (2008). *Global sensitivity analysis: The primer.* Wiley.

Schlesinger, W. H., Dietze, M. C., Jackson, R. B., Phillips, R. P., Rhoades, C. C., Rustad, L. E., & Vose, J. M. (2016). Forest biogeochemistry in response to drought. *Global Change Biology, 22,* 2318–2328. https://doi.org/10.1111/gcb.13105

Schneiderbauer, S., & Ehrlich, D., (2004). Risk, hazard and people's vulnerability to natural hazards. *European Commission Joint Research Centre. EUR, 21410.*

Serinaldi, F., Lombardo, F., & Kilsby, C. G. (2020). All in order: Distribution of serially correlated order statistics with applications to hydrological extremes. *Advances in Water Resources, 144,* 103686. https://doi.org/10.1016/j.advwatres.2020.103686

Shachter, R. D., & Kenley, C. R. (1989). Gaussian influence diagrams. *Management Science, 35,* 527–550. https://doi.org/10.1287/mnsc.35.5.527

Shen, W., Tokdar, S. T., & Ghosal, S. (2013). Adaptive Bayesian multivariate density estimation with Dirichlet mixtures. *Biometrika, 100,* 623–640. https://doi.org/10.1093/biomet/ast015

Simpson, A. H., Richardson, S. J., & Laughlin, D. C. (2016). Soil-climate interactions explain variation in foliar, stem, root and reproductive traits across temperate forests. *Global Ecology and Biogeography, 25,* 964–978. https://doi.org/10.1111/geb.12457

Simpson, E. H. (1951). The interpretation of interaction in contingency tables. *Journal of the Royal Statistical Society: Series B (Methodological), 13,* 238–241. https://doi.org/10.1111/j.2517-6161.1951.tb00088.x

Simpson, E. S., Opitz, T., & Wadsworth, J. L. (2023). High-dimensional modeling of spatial and spatio-temporal conditional extremes using INLA and Gaussian Markov random fields. *Extremes, 26,* 669–713. https://doi.org/10.1007/s10687-023-00468-8

Sivia, D., & Skilling, J. (2006). *Data analysis: A Bayesian tutorial* (2nd ed.). Oxford University Press.

Smith, R., Dick, J., Trench, H., & Van Oijen, M. (2012). Extending a Bayesian belief network for ecosystem evaluation. In *Conference Paper of the 2012 Berlin Conference of the Human Dimensions of Global Environmental Change on" Evidence for Sustainable Development", 5–6 October 2012, Berlin.*

Soyer, R. (2018). Kalman filtering and sequential Bayesian analysis. *Wiley Interdisciplinary Reviews: Computational Statistics, 10,* e1438. https://doi.org/10.1002/wics.1438

Spence, M. A., Blanchard, J. L., Rossberg, A. G., Heath, M. R., Heymans, J. J., Mackinson, S., et al. (2018). A general framework for combining ecosystem models. *Fish and Fisheries, 19,* 1031–1042. https://doi.org/10.1111/faf.12310

Spiegelhalter, D., Pearson, M., & Short, I. (2011). Visualizing Uncertainty About the Future. *Science, 333,* 1393–1400. https://doi.org/10.1126/science.1191181

Spitieris, M., & Steinsland, I. (2023). Bayesian calibration of imperfect computer models using physics-informed priors. *Journal of Machine Learning Research, 24,* 39.

Štrumbelj, E., Bouchard-Côté, A., Corander, J., Gelman, A., Rue, H., Murray, L., et al. (2024). Past, present, and future of software for Bayesian inference. *Statistical Science, Variational, 39,* 46–61.

Sutton, M. A., Simpson, D., Levy, P. E., Smith, R. I., Reis, S., van Oijen, M., & de Vries, W. (2008). Uncertainties in the relationship between atmospheric nitrogen deposition and forest carbon sequestration. *Global Change Biology, 14,* 2057–2063. https://doi.org/10.1111/j.1365-2486.2008.01636.x

ter Braak, C. J. F., & Vrugt, J. A. (2008). Differential evolution Markov chain with snooker updater and fewer chains. *Statistics and Computing, 18,* 435–446. https://doi.org/10.1007/s11222-008-9104-9

Thomas, P., Bratvold, R. B., & Eric, B. J. (2014). The risk of using risk matrices. SPE Economics and Management April 2014, D011S007R003. https://doi.org/10.2118/166269-MS

Tibshirani, R. J., Price, A., & Taylor, J. (2011). A statistician plays darts. *Journal of the Royal Statistical Society: Series A (Statistics in Society), 174,* 213–226. https://doi.org/10.1111/j.1467-985X.2010.00651.x

Tokmakian, R., Challenor, P., & Andrianakis, Y. (2012). On the use of emulators with extreme and highly nonlinear geophysical simulators. *Journal of Atmospheric and Oceanic Technology, 29,* 1704–1715. https://doi.org/10.1175/JTECH-D-11-00110.1

Tran, M.-N., Nguyen, T.-N., Dao, V.-H., 2021. A practical tutorial on Variational Bayes.

Trinh, T. H., Wu, Y., Le, Q. V., He, H., & Luong, T. (2024). Solving olympiad geometry without human demonstrations. *Nature, 625,* 476–482. https://doi.org/10.1038/s41586-023-06747-5

Van Oijen, M. (2017). Bayesian methods for quantifying and reducing uncertainty and error in forest models. *Current Forestry Reports, 3,* 269–280. https://doi.org/10.1007/s40725-017-0069-9

Van Oijen, M. (2022). *CAF2021.* https://doi.org/10.5281/zenodo.5862195

Van Oijen, M., Ågren, G. I., Chertov, O., Kellomäki, S., Komarov, A., Mobbs, D., & Murray, M. (2008). 4.4 Evaluation of past and future changes in European forest growth by means of four process-based models. *Causes and Consequences of Forest Growth Trends in Europe: Results of the Recognition Project, 21,* 183–199.

Van Oijen, M., Balkovic, J., Beer, C., Cameron, D. R., Ciais, P., Cramer, W., et al. (2014). Impact of droughts on the carbon cycle in European vegetation: A probabilistic risk analysis using six vegetation models. *Biogeosciences, 11,* 6357–6375. https://doi.org/10.5194/bg-11-6357-2014

Van Oijen, M., Beer, C., Cramer, W., Rammig, A., Reichstein, M., Rolinski, S., & Soussana, J.-F. (2013). A novel probabilistic risk analysis to determine the vulnerability of ecosystems to extreme climatic events. *Environmental Research Letters, 8,* 015032. https://doi.org/10.1088/1748-9326/8/1/015032

Van Oijen, M. & Brewer, M. (2022). *Probabilistic risk analysis and Bayesian decision theory.* SpringerBriefs in Statistics. Springer International Publishing. https://doi.org/10.1007/978-3-031-16333-3

Van Oijen, M., Cameron, D., Levy, P. E., & Preston, R. (2017). Correcting errors from spatial upscaling of nonlinear greenhouse gas flux models. *Environmental Modelling & Software, 94,* 157–165. https://doi.org/10.1016/j.envsoft.2017.03.023

Van Oijen, M., & Cameron, D. R. (2020). *BASFOR.* https://doi.org/10.5281/zenodo.3608882

Van Oijen, M., Cameron, D. R., Butterbach-Bahl, K., Farahbakhshazad, N., Jansson, P.-E., Kiese, R., et al. (2011). A Bayesian framework for model calibration, comparison and analysis: Application to four models for the biogeochemistry of a Norway spruce forest. *Agricultural and Forest Meteorology, 151,* 1609–1621. https://doi.org/10.1016/j.agrformet.2011.06.017

Van Oijen, M., Cannell, M. G. R., & Levy, P. E. (2004). Modelling biogeochemical cycles in forests: State of the art and perspectives. In F. Andersson, Y. Birot, & R. Päivinen (Eds.), *Towards the sustainable use of European forests-Forest ecosystem and landscape research: Scientific challenges and opportunities* (pp. 157–169). European Forest Institute.

Van Oijen, M., & Höglind, M. (2016). Toward a Bayesian procedure for using process-based models in plant breeding, with application to ideotype design. *Euphytica, 207,* 627–643. https://doi.org/10.1007/s10681-015-1562-5

Van Oijen, M., Höglind, M., Cameron, D. R., Thorsen, S. M., & Persson, T. (2018). *BASGRA_N.* https://doi.org/10.5281/zenodo.1493404

Van Oijen, M., Reyer, C., Bohn, F. J., Cameron, D. R., Deckmyn, G., Flechsig, M., et al. (2013). Bayesian calibration, comparison and averaging of six forest models, using data from Scots pine stands across Europe. *Forest Ecology and Management, 289,* 255–268. https://doi.org/10.1016/j.foreco.2012.09.043

Van Oijen, M., Rougier, J., & Smith, R. (2005). Bayesian calibration of process-based forest models: Bridging the gap between models and data. *Tree Physiology, 25,* 915–927. https://doi.org/10.1093/treephys/25.7.915

Van Oijen, M., & Thomson, A. (2010). Toward Bayesian uncertainty quantification for forestry models used in the United Kingdom greenhouse gas inventory for land use, land use change, and forestry. *Climatic Change, 103,* 55–67. https://doi.org/10.1007/s10584-010-9917-3

Van Oijen, M., & Zavala, M. A. (2019). Probabilistic drought risk analysis for even-aged forests. In J. Sillmann, S. Sippel, & S. Russo (Eds.), *Climate extremes and their implications for impact and risk assessment* (pp. 159–176). Elsevier.

Vaswani, A., Shazeer, N., Parmar, N., Uszkoreit, J., Jones, L., Gomez, A. N., Kaiser, L., & Polosukhin, I. (2023). *Attention Is All You Need.* https://doi.org/10.48550/arXiv.1706.03762

Vehtari, A., Gelman, A., Simpson, D., Carpenter, B., & Bürkner, P.-C. (2021). Rank-normalization, folding, and localization: An improved R^ for assessing convergence of MCMC (with discussion). *Bayesian Analysis, 16,* 667–718. https://doi.org/10.1214/20-BA1221

Welsh, A. H., Peterson, A. T., & Altmann, S. A. (1988). The fallacy of averages. *The American Naturalist, 132,* 277–288.

Wikle, C. K., & Royle, J. A. (1999). Space-time dynamic design of environmental monitoring networks. *Journal of Agricultural, Biological, and Environmental Statistics, 4,* 489–507.

Williams, C. K., & Rasmussen, C. E. (1996). Gaussian processes for regression. In *NIPS'95: Proceedings of the 8th International Conference on Neural Information Processing Systems.*

Williams, D. R., Rast, P., & Bürkner, P.-C. (2018). *Bayesian Meta-Analysis with Weakly Informative Prior Distributions.* https://doi.org/10.31234/osf.io/7tbrm

Williams, P. J., & Hooten, M. B. (2016). Combining statistical inference and decisions in ecology. *Ecological Applications, 26,* 1930–1942.

Wolfram, S. (2023). *What Is ChatGPT Doing … and Why Does It Work? Writings.*

Wong, C. (2024). How AI is improving climate forecasts. *Nature, 628,* 710–712. https://doi.org/10.1038/d41586-024-00780-8

Wood, S. (2006). *Generalized additive models: An introduction with R.* CRC Press.

Xu, Y., Choi, J., Dass, S., & Maiti, T. (2016). *Bayesian prediction and adaptive sampling algorithms for mobile sensor networks: Online environmental field reconstruction in space and time.* Springerbriefs in Control Automation and Robotics. Springer International Publishing.

Yeluripati, J. B., van Oijen, M., Wattenbach, M., Neftel, A., Ammann, A., Parton, W. J., & Smith, P. (2009). Bayesian calibration as a tool for initialising the carbon pools of dynamic soil models. *Soil Biology and Biochemistry, 41,* 2579–2583. https://doi.org/10.1016/j.soilbio.2009.08.021

Young, P. (1998). Data-based mechanistic modelling of environmental, ecological, economic and engineering systems. *Environmental Modelling & Software, 13,* 105–122. https://doi.org/10.1016/S1364-8152(98)00011-5

Zhao, W. X., Zhou, K., Li, J., Tang, T., Wang, X., Hou, Y., Min, Y., Zhang, B., Zhang, J., Dong, Z., Du, Y., Yang, C., Chen, Y., Chen, Z., Jiang, J., Ren, R., Li, Y., Tang, X., Liu, Z., … Wen, J.-R. (2023). A Survey of Large Language Models.

Index

M. van Oijen, *Bayesian Compendium*,
https://doi.org/10.1007/978-3-031-66085-6

MIX
Papier aus verantwortungsvollen Quellen
Paper from responsible sources
FSC® C105338

If you have any concerns about our products,
you can contact us on
ProductSafety@springernature.com

In case Publisher is established outside the EU,
the EU authorized representative is:
Springer Nature Customer Service Center GmbH
Europaplatz 3, 69115 Heidelberg, Germany

Printed by Libri Plureos GmbH
in Hamburg, Germany